D0164467

GASOLINE, DIESEL, AND ETHANOL BIOFUELS FROM GRASSES AND PLANTS

The world is currently faced with two significant problems – fossil fuel depletion and environmental degradation – which are continuously being exacerbated due to increasing global energy consumption. As a substitute for petroleum, renewable fuels are receiving increasing attention due to a variety of environmental, economic, and societal benefits. First-generation biofuels – ethanol from sugar or corn and biodiesel from vegetable oils – are already on the market. The goal of this book is to introduce readers to the second-generation biofuels obtained from nonfood biomass, such as forest residue, agricultural residue, switchgrass, corn stover, waste wood, and municipal solid wastes. Various technologies are discussed, including cellulosic ethanol, biomass gasification, synthesis of diesel and gasoline, biocrude by hydrothermal liquefaction, bio-oil by fast pyrolysis, and the upgradation of biofuel. This book strives to serve as a comprehensive document presenting various technological pathways and environmental and economic issues related to biofuels.

Dr. Ram B. Gupta is the PWS Distinguished Chair Professor and Chair of the Chemical Engineering Graduate Program at Auburn University. He has published numerous research papers and holds several patents on biofuels, nanotechnology, hydrogen fuel, and supercritical fluid technology and is the recipient of several national awards. He is a Fellow of the Alabama Academy of Science. He served on the editorial advisory boards of *Industrial & Engineering Chemistry Research* and *Nanomedicine: Nanotechnology, Biology and Medicine* and is currently serving on the editorial boards of *Journal of Biomedical Nanotechnology, Research Letters in Nanotechnology, Open Nanomedicine Journal, International Journal of Chemical Engineering*, and *Research Letters in Chemical Engineering*. His recent books are *Nanoparticle Technology for Drug Delivery, Solubility in Supercritical Carbon Dioxide*, and *Hydrogen Fuel: Production, Transport, and Storage*.

Dr. Ayhan Demirbas is Professor and Vice Rector at Sirnak University. His research on renewable and sustainable energy has been published in 445 scientific papers. He served on the editorial advisory board of *Energy Conversion and Management* and is currently serving as the Editor-in-Chief of *Energy Education Science and Technology Part A: Energy Science and Research, Energy Education Science and Technology Part B: Social and Educational Studies, Future Energy Sources*, and *Social Political Economic and Cultural Research*. His recent books are *Biodiesel: A Realistic Fuel Alternative for Diesel Engines, Biofuels: Securing the Planet's Future Energy Needs, Biohydrogen: For Future Engine Fuel Demands, Biorefineries: For Biomass Upgrading Facilities, Methane Gas Hydrate*, and *Algae Energy*.

Gasoline, Diesel, and Ethanol Biofuels from Grasses and Plants

Ram B. Gupta

Auburn University

Ayhan Demirbas

Sirnak University

CAMBRIDGE UNIVERSITY PRESS
Cambridge, New York, Melbourne, Madrid, Cape Town,
Singapore, São Paulo, Delhi, Tokyo, Mexico City

Cambridge University Press
32 Avenue of the Americas, New York, NY 10013-2473, USA

www.cambridge.org
Information on this title: www.cambridge.org/9780521763998

First published 2010
Reprinted 2011

A catalog record for this publication is available from the British Library.

Library of Congress Cataloging in Publication Data

Gupta, Ram B.
Gasoline, diesel, and ethanol biofuels from grasses and plants / Ram B. Gupta,
Ayhan Demirbas.
    p.   cm.
Includes bibliographical references and index.
ISBN 978-0-521-76399-8 (hardback)
1. Plant biomass. 2. Forest biomass. 3. Biomass energy. I. Demirbas,
Ayhan. II. Title.
TP248.27.P55G87   2010
662'.88–dc22   2009042276

ISBN 978-0-521-76399-8 Hardback

Contents

Preface

The world is currently faced with two significant problems: fossil fuel depletion and environmental degradation. The problems are continuously being exacerbated due to increasing global population and per capita energy consumption. To overcome the problems, renewable energy has been receiving increasing attention due to a variety of environmental, economic, and societal benefits. First-generation biofuels (ethanol from sugar or corn, and biodiesel from vegetable oils) are already in the market, and second-generation biofuels from nonfood biomass are under development. The goal of this book is to introduce readers to the biofuels obtained from nonfood biomass, and for reference to provide the technologies involved in first-generation biofuels derived from food sources.

Chapter 1 discusses various nonrenewable (petroleum, natural gas, coal) and renewable forms of energy, and describes air pollution and greenhouse gas emission caused by the use of fossil fuels. Recent concern about carbon dioxide emissions, carbon sequestration, and carbon credits are discussed in Chapter 2. Chapter 3 provides an in-depth description of various renewable energy sources, including biomass; hydropower; geothermal, wind, solar, and ocean energy; and biogas. For the production of biofuels, the global availability of biomass is discussed in Chapter 4 along with the characterization and variations of biomass.

Conventional ethanol production from corn or sugarcane by fermentation technology is discussed in Chapter 5. Current techniques and various unit operations involved are presented, including saccharification, fermentation, distillation, and dehydration. The second-generation ethanol from cellulose is described in Chapter 6. It provides an in-depth coverage of various pretreatment techniques that are critical to the cost-effective production of cellulosic ethanol. In addition, xylose fermentation to improve the ethanol yield is discussed.

Chapter 7 discusses the production of biodiesel from vegetable oil by transesterification. The fuel properties of biodiesel are compared with those of petroleum diesel. Chapter 8 concerns the production of diesel from biomass. Processing of biomass gasification followed by Fischer–Tropsch synthesis of diesel and other liquid fuels is discussed. Chapter 9 outlines the production of bio-oil from biomass by the pyrolysis process. Various reactor designs for fast pyrolysis are described along with the fuel properties of bio-oil, including its upgradation. Chapter 10 deals with the production of biocrude by hydrothermal liquefaction of biomass, in which various aspects of production and upgradation are presented to obtain fuel comparable to petroleum liquids.

Chapter 11 discusses the use of wind and solar energies to enhance biofuel production from biomass. The process heating and electricity needs can be satisfied so that a higher amount of biomass carbon is converted to liquid fuels. Chapters 12 and 13 discuss the environmental and economic impacts of biofuels, respectively. Chapter 14 summarizes current biofuel policies of major countries that are promoting biofuel production and use.

This book strives to serve as a comprehensive document to present various technological pathways and environmental and economic issues related to biofuels. As petroleum reserves are depleted, the world is faced with finding alternatives. Currently, the transport sector depends almost entirely on petroleum liquids (diesel, gasoline, jet fuel, kerosene), and to fill the gap, biofuel can provide a replacement. However, alternatives to petroleum must be technically feasible, economically competitive, environmentally acceptable, and easily available.

The authors are thankful for assistance from various people in preparation of this manuscript, including Mr. Sandeep Kumar, Mrs. Sweta Kumari, Dr. Lingzhao Kong, Mrs. Hema Ramsurn, and Prof. Sushil Adhikari. In addition, support from our families – Deepti, Pranjal, and Rohan Gupta; Elmas, Temucin, Kursat, Muhammet, Ayse Hilal, and Burak Demirbas – was key to the completion of this book.

Ram B. Gupta, Auburn, USA
Ayhan Demirbas, Trabzon, Turkey

1 Introduction

1.1 Energy

Energy is defined as the ability to do work and provide heat. There are different ways in which the abundance of energy around us can be stored, converted, and amplified for our use. Energy sources can be classified into three groups: fossil, renewable, and nuclear (fissile). Fossil fuels were formed in an earlier geological period and are not renewable. The fossil energy sources include petroleum, coal, bitumen, natural gas, oil shale, and tar sands. The renewable energy sources include biomass, hydro, wind, solar (both thermal and photovoltaic), geothermal, and marine. The main fissile energy sources are uranium and thorium. Despite adequate reserves, some classifications include fissile materials along with the nonrenewable sources.

For over ten thousand years, humans have used biomass for their energy needs. Wood was used for cooking, water, and space heating. The first renewable energy technologies were primarily simple mechanical applications and did not reach high energetic efficiencies. Renewable energies have been the primary energy source in the history of the human race. But in the last two hundred years, we have shifted our energy consumption toward fossil fuels. Industrialization changed the primary energy use from renewable resources to sources with a much higher energy density, such as coal or petroleum. During the last century, the promise of unlimited fossil fuels was much more attractive, and rapid technical progress made the industrial use of petroleum and coal economical.

Petroleum is the largest single source of energy consumed by the world's population (about 4.8 barrel/year/person), exceeding coal, natural gas, nuclear, or hydroelelctric, as shown in Figure 1.1. The United States' energy consumption and supply is shown in Figure 1.2. In fact, today, over 88% of the global energy used comes from three fossil fuels: petroleum, coal, and natural gas. Although fossil fuels are still being created today by underground heat and pressure, they are being consumed much more rapidly than they are created. Hence, fossil fuels are considered nonrenewable, that is, they are not replaced as fast as consumed. Unfortunately, petroleum oil is in danger of becoming short in supply. Hence, the future trend is

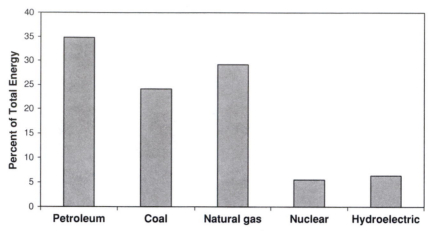

Figure 1.1. World consumption of various energies in 2008 out of the total consumption of 11.3 billion tons oil equivalent (BP, 2009).

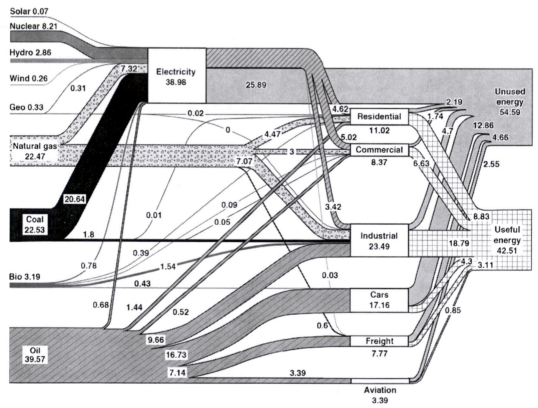

Figure 1.2. Energy consumption and supply in the United States for 2006 (U.S. DOE, 2008). Energies are shown in quad units (1 quad = 10^{15} BTU = 1.055×10^{15} J).

toward using alternate energy sources. Fortunately, technological development is making the transition possible.

A major problem with petroleum fuels is their uneven distribution in the world; for example, about 2% of the world population in the Middle East has 63% of the global reserves and is the dominant supplier of petroleum. This energy system is unsustainable because of equity issues as well as environmental, economic, and geopolitical concerns that have far-reaching implications. Interestingly, renewable energy resources are more evenly distributed than fossil or nuclear resources. Also, the energy flows from renewable resources are more than three orders of magnitude higher than current global energy need. Hence, renewable energy sources, such as biomass, hydro, wind, solar (both thermal and photovoltaic), geothermal, and marine, will play an important role in the world's future supply. For example, it is estimated that by year 2040 approximately half of the global energy supply will come from renewables (EREC, 2006), and the electricity generation from renewables will be more than 80% of the total global electricity production.

Another major problem with fossil fuel is the greenhouse gas emissions; about 98% of carbon emissions result from fossil fuel combustion. Reducing the use of fossil fuels would considerably reduce the amount of carbon dioxide and other pollutants produced. This can be achieved by either using less energy altogether or by replacing fossil fuels with renewable fuels. Hence, current efforts focus on advancing technologies that emit less carbon (e.g., high-efficiency combustion) or no carbon, such as nuclear, hydrogen, solar, wind, and geothermal, or on using energy more efficiently and sequestering carbon dioxide that is emitted during fossil fuel combustion.

Despite the above challenges, it is not easy to replace fossil fuels, as our modern way of life is intimately dependent on fossil fuels, specifically hydrocarbons, including petroleum, coal, and natural gas. For example, the majority of commodity products (e.g., plastics, fabrics, machine parts, chemicals, etc.) are made using crude oil or natural gas feedstock. And, more importantly, the major energy demand is fulfilled by fossil fuels, resulting in a major role for crude oil and natural gas in driving world economy.

1.2 Petroleum

Petroleum [word derived from Greek *petra* (rock) and *elaion* (oil) *or* Latin *oleum* (oil)] or crude oil, sometimes colloquially called black gold or "Texas Tea," is a thick, dark brown or greenish liquid. Petroleum consists of a complex mixture of various hydrocarbons, largely of the alkenes and aromatic compounds, but may vary much in appearance and composition. Petroleum is a fossil fuel because it was formed from the remains of tiny sea plants and animals that died millions of years ago and sank to the bottom of the oceans (Figure 1.3). This organic mixture was subjected to enormous hydraulic pressure and geothermal heat. Over time, the mixture changed, breaking down into compounds made of hydrocarbons by reduction

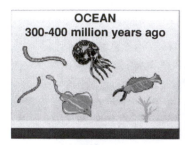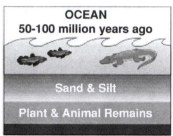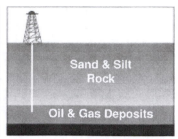

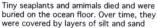

| Tiny seaplants and amimals died and were buried on the ocean floor. Over time, they were covered by layers of silt and sand | Over millions of years, the remains were buried deeper and deeper. The enormous heat and pressure turn them into oil and gas | Today, we drill down through layers of sand, silt, and rock to rich the rock formation that contains oil and gas deposits |

Figure 1.3. Cartoon showing formation of petroleum and natural gas (U.S. DOE, 2009).

reactions. This results in the formation of oil-saturated rocks. The crude oil rises and gets trapped under nonporous rocks that are sealed with salt or clay layers.

According to the well-accepted Biogenic theory (Walters, 2006), fossil fuels – crude oil, coal, and natural gas – are the product of compression and heating of ancient vegetation and animal remains over geological time scales. According to this theory, an organic matter is formed from the decayed remains of prehistoric marine animals and terrestrial plants. Over many centuries, this organic matter, mixed with mud, is buried under thick sedimentary layers. The resulting high pressure and heat transformed the organic matter first into a waxy material known as kerogen, and then into liquid and gaseous hydrocarbons. The fluids then migrate through adjacent rock layers until they become trapped underground in porous rocks called reservoirs, forming an oil field, from which the liquid can be removed by drilling and pumping. The reservoirs are at different depths in different parts of the world, but typical depth is 4–5 km. The thickness of the oil layer is about 150 meters, generally termed "oil window." Three important elements of an oil reservoir are a rich source rock, a migration conduit, and a trap (seal) that forms the reservoir.

According to not-well-accepted Abiogenic theory (Mehtiev, 1986), petroleum origin is natural hydrocarbons. This theory proposes that large amounts of carbon exist naturally in the planet, some in the form of hydrocarbons. Due to its lower density than aqueous pores fluids, hydrocarbons migrate upward through deep fracture networks. The two theories are reviewed by Mehtiev (1986).

1.2.1 History of Petroleum Exploration

According to historical accounts, the early oil wells were drilled in China before the fifth century (ACE, 2009). Wells, as deep as 243 meters, were drilled using bits attached to bamboo poles. The crude oil was burned to produce heat needed in the production of salt from brine evaporation. By the end of the tenth century, extensive bamboo pipelines connected oil wells with salt springs.

Separately, ancient Persian tablets indicate the medicinal and lighting uses of petroleum in the upper echelons of their society. Tar, the heavy component of the oil, was used in the paving of the street in Baghdad in the eighth century. In Baku

(Azerbaijan), oil fields were exploited to produce naphtha in the ninth century, as described by the geographer Masudi in the tenth century, and the output has increased to hundreds of shiploads by the thirteenth century as described by Marco Polo.

Modern petroleum refining began in 1846 when Abraham Gesner refined kerosene from coal. Six years later, Ignacy Łukasiewicz refined kerosene from "rock oil," which was more readily available. The successful process rapidly spread around the world. In 1861, Meerzoeff built the first Russian refinery in Baku (current capital of the Azerbaijan Republic), using the mature oil fields available. It is interesting to note that the battle of Stalingrad was fought over Baku, which produced 90% of the world's oil.

The first commercial oil well in North America was drilled by James Miller Williams in 1858 in Oil Springs (Ontario, Canada). In the following year, Edwin Drake discovered crude oil near Titusville, Pennsylvania and pioneered a new method for producing crude oil from the ground, in which the drilling was carried out using piping to prevent borehole collapse, allowing for a deeper drilling. The new method was significant advancement, as the previous methods for collecting crude oil were very limited. For example, the collection used to be performed where crude oil occurred naturally, such as oil seeps or shallow holes dug into the ground. Also, due to the collapse from water seepage, the digging of large shafts into the ground almost always failed. A notable advancement Drake achieved was to place a 10-meter iron pipe through the ground all the way to the bedrock below. This allowed Drake to do further drilling inside the pipe, avoiding the hole collapse from water seepage. Typical drilling below the pipe was to 13 meters, making a 23-meter well. The basic concept in Drake's technology is still being used by many petroleum companies, with the new wells being 1,000–4,000 meters deep. While searching for oil, many of the test wells turn up dry (i.e., do not contain any oil), which adds to the cost of petroleum exploration. Hence, some small investors still consider the exploration as a gamble. Despite the improvement in the technology since Drake's time, only about 33% of the exploration wells have oil; the remaining 67% come up dry.

In American homes and businesses, whale oil was used for lighting until 1850s and often experienced shortages. Starting in the 1860s, refined kerosene from crude oil became plentiful; hence, kerosene was used for lighting until the discovery of electric bulbs. Gasoline and other non-kerosene products from refining were simply discarded due to the lack of use. In 1882, the advent of gasoline engine-driven carriages (i.e., horseless carriages) solved this problem. There was heavy demand for gasoline; in fact, by 1920, there were 9 million motor vehicles in the United States alone for a population of 106 million.

1.2.2 Petroleum Refining and Shipping

A petroleum refinery separates crude oil into various byproducts and fuel components, including gasoline, diesel, heating oil, and jet fuel. Petroleum constituents

Table 1.1. *Petroleum constituents from oil refinery*

Fraction	Distillation range (°C)	Carbon number
Gas	<20	C_1–C_4
Petroleum ether	20–60	C_5–C_6
Ligroin (light naphtha)	60–100	C_6–C_7
Natural gasoline	40–205	C_5–C_{10}, and cycloalkanes
Jet fuel	105–265	C_8–C_{14}, and aromatics
Kerosene	175–315	C_{10}–C_{16}, and aromatics
No. 1 diesel fuel	170–325	C_9–C_{18}, and aromatics
No. 2 diesel fuel	175–365	C_{10}–C_{20}, and aromatics
No. 3 diesel fuel	185–390	C_{12}–C_{24}, and aromatics
Gas oil (No. 4 and 5 fuel oils)	>275	C_{12}–C_{70}, and aromatics
Gas oil (No. 6 fuel oils)	>365	C_{20}–C_{70}, and aromatics
Lubricating oil	Nonvolatile liquids	Long chains attached to cyclic structures
Asphalt or petroleum coke	Nonvolatile solids	Polycyclic structures

from oil refinery products are listed in Table 1.1. Because various components boil at different temperatures (or temperature ranges), refineries use a heating process called distillation to separate the components. For example, gasoline has a lower boiling temperature (i.e., gasoline is more volatile) than kerosene, allowing the two to be separated by heating to different temperatures. Another important job of the refineries is to remove contaminants from the oil, for example, sulfur from gasoline or diesel. If not removed, the sulfur will be emitted as sulfur dioxide from automobile exhaust, causing air pollution and acid rain.

An important nonfuel use of petroleum is to produce chemical raw materials. The two main classes of petrochemical raw materials are olefins (including ethylene and propylene) and aromatics (including benzene and xylene), both of which are produced in large quantities. The olefins are produced by chemical cracking by using steam and/or catalysts, and the aromatics are produced by catalytic reforming. These two basic types of chemicals serve as feedstock to produce a wide range of chemicals and materials, including monomers, solvents, and adhesives. The monomers are used to produce polymers and oligomers for use as plastics, resins, fibers, elastomers, lubricants, and gels. An important aspect of petrochemicals is their extremely large scale. For example, annual world production of ethylene is 110 million tons, propylene is 65 million tons, and aromatics are 70 million tons. Due to the economy of scale, each plant produces a large volume of petrochemicals, which are then distributed worldwide via interregional trade. The majority of the petrochemical industry is in the United States and Western Europe, although the production capacity in the Middle East and Asia is increasing due to the proximity of raw materials and end consumers.

Each day, Americans use about 21 million barrels of crude oil, out of which about one-third is produced domestically and two-thirds are imported. Most U.S. production is in Texas followed by Alaska, California, Louisiana, and Oklahoma, in that order. Gasoline and other liquid fuels from refineries are usually shipped out through pipelines to the major consumer centers, which is the safest and cheapest way to move large quantities of petroleum across land. The United States has an extensive network of underground pipelines measuring 384,000 km. Pump stations are installed at a spacing of 30–170 km to keep the petroleum liquids flowing at a speed of about 8 km/hour. For illustration purposes, it takes about 15 days to send a shipment of gasoline from Houston to New York.

Petroleum is the most important commodity in terms of international trade. Hence, it has been subjected to influence from the international groups. A notable group is the Organization of Petroleum Exporting Countries (OPEC), which is an intergovernmental organization of 13 nations: Iran, Iraq, Kuwait, Saudi Arabia, Venezuela, Qatar, Indonesia, Libya, United Arab Emirates, Algeria, Nigeria, Ecuador, and Gabon. OPEC members try to set production levels for petroleum to maximize their revenue by using supply/demand economics. For example, a decrease in the production level increases the price ($/barrel), and on the contrary, an increase in the production level lowers the price. But the total money a country receives is barrel of crude oil produced × price per barrel. Due to this complex dependence of revenue on production volume, member OPEC countries do not always agree with each other; as depending on the production levels for a member country, the optimum price may be different from another member. Hence, contrary to the group's vision, some OPEC countries want to produce less to raise prices, whereas the other OPEC countries want to flood the market to gain immediate revenue. In addition, the crude oil supply can be controlled for political reasons. For example, the 1973 OPEC oil embargo was a political statement against the United States for supporting Israel in the Yom Kippur war. Such embargo or cut in production caused a drastic increase in the price of petroleum. Today, a significant portion of U.S. crude oil import is from Canada and Mexico, which is more reliable and has a lower shipping cost. However, due to an internal law, Mexico can only export half of the petroleum it produces to the United States.

The United States is a member of the Organization for Economic Co-Operation and Development (OECD), which is an international organization of 30 countries that accept the principles of representative democracy and a free market economy. In 1970s, as a counterweight to OPEC, OECD founded the International Energy Agency (IEA), which is regarded as the "energy watchdog" of the Western world and is supposed to help avoid future petroleum crises. IEA provides demand and supply forecasts in its annual World Energy Outlook (WEO) report and current situation of crude oil market in its monthly publication. The WEO covers forecast for the next two decades and is highly regarded by people related to the energy industry.

1.2.3 Classification of Oils

The oil industry classifies "crude" by its production location (e.g., "West Texas Intermediate, WTI" or "Brent"), relative density (API gravity), viscosity ("light," "intermediate," or "heavy"), and sulfur content ("sweet" for low sulfur and "sour"

Table 1.2. *Class of compounds found in petroleum crude oils (Robins and Hsu, 2000)*

Compounds	Molecular structures

for high sulfur). The density and viscosity of the crude oil depend on the composition of the oil, which varies greatly from one source to another, and the age of the producing well. Some of the class of compounds are listed in Table 1.2.

Tar sands are oil traps that are not deep enough in the earth to allow for geological conversion into conventional oil. This oil was not heated enough to complete the process of molecular breakage to reduce the viscosity. The oil has the characteristics of bitumen and is mixed with large amounts of sand due to the proximity to the earth's surface. The tar sand is mined, flooded with water in order to separate the heavier sand, and then processed in special refineries to reduce its high sulfur content (the original crude oil usually has 3–5% sulfur) and other components. This process needs huge amounts of energy and water. The deeper tar sands (below 75 meters) are mined in situ (COSO, 2007).

Natural gas liquids are liquid hydrocarbons produced along with natural gas (NG) from the oil wells. The chemical composition of NG is given in Table 1.3.

Oil shales contain kerogen, which is an intermediate stage on the way from biological carbohydrate to oil formation. The oil shale layer had not been subjected to enough heat in order to complete the conversion. For the final step, the kerogen must be heated to 500°C and molecularly combined with additional hydrogen to

Table 1.3. *Typical chemical composition of natural gas*

Component	Typical analysis (v/v%)	Range (v/v%)
Methane	94.9	87.0–96.0
Ethane	2.5	1.8–5.1
Propane	0.2	0.1–1.5
i-Butane	0.03	0.01–0.3
n-Butane	0.03	0.01–0.3
i-Pentane	0.01	trace–0.14
n-Pentane	0.01	trace–0.14
Hexanes plus	0.01	trace–0.06
Nitrogen	1.6	1.3–5.6
Carbon dioxide	0.7	0.1–1.0
Oxygen	0.02	0.01–0.1
Hydrogen	trace	trace–0.02

complete the oil formation. The final processing is done in the refinery and requires high amounts of energy (otherwise the energy would have come from the geological environment). The kerogen is still embedded in the source rock and did not concentrate as in the case of crude oil field. Often it is not attractive to mine the kerogen, due to the large amount of waste rock that comes along. However, the shale oil reserves in the world are greater than crude oil or NG, as shown in Table 1.4. Researchers are devising innovative ways to mine kerogen in situ by means of heat and steam.

1.2.4 Petroleum Reserves and Crude Oil Production

The petroleum reserves can be classified into three categories: proven, probable, and possible reserves. Proven reserves are those fields from which petroleum can be produced using current technology at the current prices. Probable reserves are those fields from which petroleum can be produced using the near-future technology at current prices. Possible reserves are those fields from which petroleum can be produced using future technology. Relying on our proven reserves, currently petroleum is the most important energy source, as 35% of the world's primary energy need is met by crude oil, 25% by coal, and 21% by NG, as shown in Table 1.5. The transport sector (i.e., automobiles, ships, and aircrafts) relies on well over 90% of crude oil. In fact, the economy and lifestyle of industrialized nations rely heavily on a sufficient supply of crude oil at low cost.

Table 1.6 shows crude oil production data for various regions. The Middle East produces 32% of the world's oil (Table 1.7), but more importantly, it has 64% of the total proven oil reserves in the world (Table 1.8). Also, the Middle East's reserves are depleting at a slower rate than any other region in the world. The Middle East provides more than half of OPEC's total oil exports and has a major influence on the worldwide crude oil prices, despite the fact that OPEC produces less than half of the oil in the world.

Table 1.4. *Energy reserves of the world (Demirbas, 2006a; Demirbas, 2006b)*[a]

Crude oil	Natural gas	Shale oil	Coal	Tar sands	Uranium	Deuterium
37.0	19.6	79.0	320.0	6.1	1.2×10^5	7.5×10^9

[a] Each unit $= 1 \times 10^{15}$ MJ $= 1.67 \times 10^{11}$ Bbl crude oil.

Table 1.5. *Year 1973 and 2005 fuel shares of total primary energy supply (excludes electricity and heat trade) (IEA, 2007)*

Energy sources	World		OECD	
	1973	2005	1973	2005
Oil	46.2	35.0	53.0	40.6
Coal	24.4	25.3	22.4	20.4
Natural gas	16.0	20.7	18.8	21.8
Combustible renewables and wastes	10.6	10.0	2.3	3.5
Nuclear	0.9	6.3	1.3	11.0
Hydro	1.8	2.2	2.1	2.0
Other (geothermal, solar, wind, heat, etc.)	0.1	0.5	0.1	0.7
Total (million tons oil equivalent)	6,128	11,435	3,762	5,546

Table 1.6. *Year 1973 and 2006 regional shares of crude oil production (IEA, 2007)*

Region	1973	2006
Middle East (%)	37.0	31.1
OECD (%)	23.6	23.2
Former USSR (%)	15.0	15.2
Africa (%)	10.0	12.1
Latin America (%)	8.6	9.0
Asia, excluding China (%)	3.2	4.5
China (%)	1.9	4.7
Non-OECD Europe (%)	0.7	0.2
Total (million tons)	2,867	3,936

Table 1.7. *Percentage of petroleum production by region (IEA, 2007)*

Middle East	Latin America	Eastern Europe	North America	Asia and Pacific	Africa	Western Europe
32	14	13	11	11	10	9

Table 1.8. *Percentage of total proven reserves by region (IEA, 2007)*

Middle East	Latin America	Eastern Europe	North America	Asia and Pacific	Africa	Western Europe
64	12	6	3	4	9	2

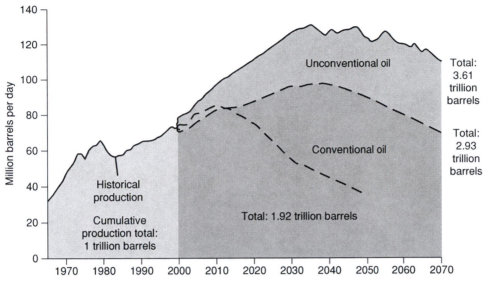

Figure 1.4. Crude oil production scenario. Twin peaks: peak-oil supporters think we have already reached or will soon reach a historical maximum of oil production; others argue that oil production will not peak until at least 2030 (Witze, 2007; reproduced with permission from Nature Publishing Group; original data from Cambridge Energy Research Associates).

At the present rate of petroleum production, the smaller petroleum reserves are on the verge of depletion, and the larger reserves are estimated to be depleted in less than 50 years. Hence, the world is facing a bleak future of petroleum short supply. Figure 1.4 shows global crude oil production scenarios based on today's production. According to one estimate, peak in global oil production is likely to occur by 2012, and thereafter the production will start to decline at a rate of several percent per year. By 2030, the global petroleum supply will be dramatically lower, which will create a supply gap that may be hard to fill by growing contributions from other fossil, nuclear, or alternative energy sources in that time frame. However, there are other estimates that the peak production will occur beyond 2030.

1.2.5 Crude Oil Pricing

The price of a barrel (42 gallons or 159 liters) of crude oil is highly dependent on both its grade (e.g., specific gravity, sulfur content, viscosity) and location. References to the oil prices are usually references to the spot price of either WTI/Light Crude as traded on New York Mercantile Exchange (NYMEX) for delivery in Cushing (Oklahoma) or the price of Brent as traded on the International Commodities Exchange (ICE) for delivery at Sullom Voe (Shetland, Scotland). The price is highly influenced by the demand and supply (both current and perceived future supplies), which in turn are highly dependent on global macroeconomic and political conditions. It is often claimed that OPEC sets a high crude oil price and the true cost of crude oil production is only $2/barrel in the Middle East.

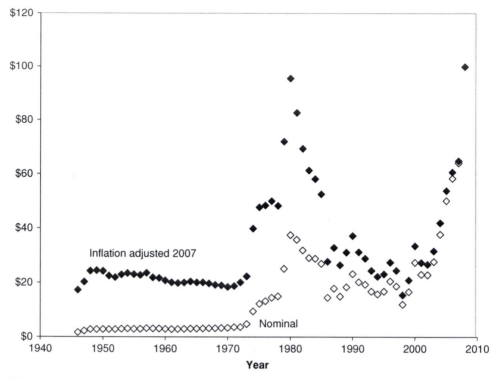

Figure 1.5. Historical per barrel crude oil prices for US refineries (EIA, 2009a).

These estimates are based only on the operating cost of running a well and do not include the cost of finding and developing an oil field. Based on the macroeconomics, the price of crude oil is based on what it cost to produce the next barrel of crude oil over the current level. Hence, the price set by not easy-to-produce but that by difficult-to-produce marginal oil. For example, the production limits imposed by OPEC have caused the development of North Sea oil fields, where it is more expensive to produce the oil. On the other hand, investing in spare capacity is expensive, and the low oil price environment of the 1990s has led to cutbacks in the investment. As a result, during the oil price rally seen since 2003, OPEC's spare capacity has not been sufficient to stabilize prices, as shown in the Figure 1.5. The peak crude oil price in 2008 of $145/barrel put a severe strain on many world economies, which in part contributed to the recession that followed.

1.3 Natural Gas

Usefulness of NG has been known for hundreds of years, as it has been used to heat water and light lamps. In recent years, NG has become the fastest growing primary energy source in the world, mainly because it is a cleaner fuel than crude oil or coal and is not as controversial as nuclear power. NG combustion is clean and emits less carbon dioxide (CO_2) than all the other petroleum fuels, making it

Table 1.9. *World natural gas reserves by country (Demirbas, 2008a)*

Country	Reserves (trillion cubic meter)	% of world total
Russian Federation	48.1	33.0
Iran	23.0	15.8
Qatar	8.5	5.8
United Arab Emirates	6.0	4.1
Saudi Arabia	5.8	4.0
United States	4.7	3.3
Venezuela	4.0	2.8
Algeria	3.7	2.5
Nigeria	3.5	2.4
Iraq	3.1	2.1
Turkmenistan	2.9	2.0
Malaysia	2.3	1.6
Indonesia	2.0	1.4
Uzbekistan	1.9	1.3
Kazakhstan	1.8	1.3
Rest of world	23.8	16.5

favorable in terms of global warming. NG is used in various sectors, including industrial, residential, electric generation, commercial, and transportation. NG is found in many parts of the world, but the largest reserves are in Russia and the Middle East (Table 1.9). Since the 1970s, world NG reserves have generally increased each year.

Worldwide, NG use is steadily increasing (Figure 1.6) due to a variety of reasons, including price ($/GJ), environmental concerns (reduced emissions and global warming), fuel diversification for energy security, market deregulation, and overall economic growth.

In NG consumption, the United States ranks first, Russia ranks second, and Europe ranks third. The largest NG producer is Russia, which is also the largest supplier of NG to Western Europe; the Russian supply to Europe and to the other

Figure 1.6. Production of natural gas in the world (BP, 2009).

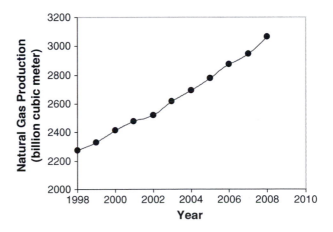

former Soviet Republic is influenced by political motives. Both Asia and Oceania import NG from other regions to satisfy their demands.

Compared with oil, only a moderate amount of NG is traded on the world markets, mainly because of the low density of NG, making it more expensive to transport than oil. For example, a given pipeline can transport 15-fold more energy as liquid oil when compared with transporting high-pressure NG. Hence, the gas pipelines need to have much larger diameter and/or fluid velocity for a given energy movement.

In fact, pipeline transportation is not always feasible because of the growing geographic distance between gas reserves and the consumers. Many of the importing countries do not wish to solely rely on NG import due to the potential political instabilities that may affect the long pipeline routes. The alternate transport routes are by ships or railcars. But, for economical transport, sufficient energy needs to be packaged in the containers, which is done by liquefaction. A full liquefied natural gas (LNG) chain consists of (1) a liquefaction plant, (2) low temperature, pressurized, transport ships, (3) and a regasification terminal at the destination. All three steps are a significant amount of energy. Hence, it is not surprising to see that, as opposed to worldwide consumption of 2.4 billion tons of NG, only 60 million tons of LNG is traded (of which 39 million tons is imported by Japan).

The generation of electricity is an important use of NG. However, the electricity from NG is generally more expensive than using coal because of increased fuel costs. NG can be used to generate electricity in a variety of ways: (1) conventional steam generation, similar to coal-fired power plants in which heating is used to generate steam, which in turns runs turbine with an efficiency of 30–35%; (2) centralized gas turbine, in which hot gases from NG combustion are used to turn the turbine; and (3) combine cycle unit, in which both steam and hot combustion gases are used to turn the turbine with an efficiency of 50–60%.

The use of NG in power production has increased due to the fact that NG is the cleanest-burning alternative fossil fuel. Upon combustion of fossil fuel, both carbon and hydrogen atoms combine with oxygen to form CO_2 and water (H_2O) along with heat. NG (methane, CH_4) contains less C and more H as compared with coal or petroleum; hence, less CO_2 and more H_2O is produced upon combustion. In addition, NG combustion produced virtually no sulfur dioxide (SO_2), and only a small amount of nitrous oxides (NO_x). Concerns about acid rain, urban air pollution, and global warming are likely to increase NG use in the future. NG burns far cleaner than gasoline or diesel, producing fewer NO_x, unburned hydrocarbons, and particulates. Because NG vehicles require large storage tanks, the main market may be for buses for use within the cities. Another use that may develop is the use of fuel cell for stationary and transportation application. The energy for fuel cells comes from hydrogen, which can be produced from NG. With a very few moving parts and operation at low temperatures (typically 70°C), fuel cells can provide a high fuel efficiency (typically 60%). In addition, the low temperature operation avoids the formation of NO_x pollutants, which are formed during conventional high temperature combustion of any fuel.

Table 1.10. *Worldwide amounts of organic carbon sources (Hacisalihoglu, Demirbas, and Hacisalihoglu, 2008)*

Source of organic carbon	Amount (gigaton)
Gas hydrates (onshore and offshore)	10,000–11,000
Recoverable and nonrecoverable fossil fuels (oil, coal, gas)	5,000
Soil	1,400
Dissolved organic matter	980
Land biota	880
Peat	500
Others	70

1.3.1 Methane from Gas Hydrates

Methane gas hydrates are crystalline solids formed by combination of CH_4 and H_2O at low temperatures and high pressures. Gas hydrates have an ice-like crystalline lattice of water molecules with methane molecules trapped inside. Enormous reserves of hydrates can be found under continental shelves and on land under permafrost. The amount of organic carbon in gas hydrates is estimated to be twice that in all other fossil fuels combined (Table 1.10). However, due to the solid form of gas hydrates, conventional gas and oil recovery techniques are not suitable. The recovery of CH_4 generally involves dissociating or melting in situ gas hydrates by heating the reservoir above the temperature of hydrate formation or by decreasing the reservoir pressure below hydrate equilibrium (Lee and Holder, 2001).

1.4 Coal

Coal is the oldest fossil fuel known. Coal is a heterogeneous combustible material containing components that have fuel value (carbon and hydrogen), components with no fuel value (moisture, ash, oxygen), and pollutants (sulfur, mercury, etc.). Coal is formed from the decomposition of plant matters over millions of years. When plants die in a low-oxygen swamp environment, instead of decaying by bacteria and oxidation, their organic matter is preserved (Figure 1.7). Over time, heat and pressure remove the H_2O and transform the matter into coal via several stages. The

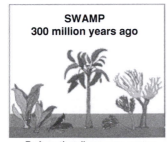
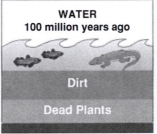

| Before the dinosaurs, many giant palnts died in swamps | Over million of years, the plants were buried under water and dirt | Heat and pressure turned The dead plants into coal |

Figure 1.7. Cartoon showing formation of coal (U.S. DOE, 2009).

Table 1.11. *Chemical properties of typical coal samples*

	Low-rank coal	High-volatile coal	High-rank coal
Carbon, %	75.2	82.5	90.5
Hydrogen, %	6.0	5.5	4.5
Oxygen, %	17.0	9.6	2.6
Nitrogen, %	1.2	1.7	1.9
Sulfur, %	0.6	0.7	0.5
Moisture, %	10.8	7.8	6.5
Calorific value, MJ/kg	31.4	35.0	36.0

first stage in coal formation yields peat, which is a compressed plant matter containing leaves and twigs. The second stage yields brown coal or lignite, which contains much lower moisture, oxygen, and nitrogen as compared with peat. The third stage yields bituminous coal. The fourth stage yields anthracite coal, which is harder than bituminous coal. Hence, coal is classified into four general categories: lignite (young coal) through sub-bituminous and bituminous to anthracite.

Lignite is widely used as a heating fuel but is of little chemical interest. Bituminous is the most abundant form of coal and is used for heating and producing chemicals and materials (e.g., coke for smelting). The chemical composition (wt%) and calorific values of various coals are given in Table 1.11.

The worldwide coal production was 6.5 billion tons in year 2007. With the estimated reserve of 909 billion tons (recoverable coal), the coal is expected to last for another 140 years at the current consumption rate. The major world deposits of coal are in the United States (27.1%), Russia (17.3%), China (12.6%), India (10.2%), Australia (8.6%), and South Africa (5.4%). Although coal is much more abundant than crude oil or gas on a global scale, coalfields can easily get depleted on a regional scale. Due to its abundance and wide distribution, coal accounts for 25% of the world's primary energy consumption and 37% for electricity generation. The coal can be cost-effectively mined, transported, and stored by the use of modern techniques. In addition, the international trade has kept the prices low by vigorous competition. But now, the growth in coal's future use depends on environmental performance, particularly the efforts of the power-generation industry in reducing the pollutant emissions, including CO_2. The nonfuel use of coal includes production of metallurgical coke, activated carbon, specialty aromatic and aliphatic chemicals, methanol, and hydrogen.

1.5 Biofuels

First-generation biofuels – bioethanol and biodiesel – are derived from carbohydrates (e.g., grains, sugar) and vegetable oils, respectively. Second-generation biofuels are derived form biomass (lignocellulosic materials), including wood, stalks, and grasses. Together, these represent existing and potential biomass resources. The share of biofuel in the automotive fuel market is likely to grow rapidly in the next decade due to the numerous benefits, including sustainability, reduction of greenhouse gas emissions, regional development, reduction of rural poverty, and fuel security.

The biofuel economy, and its associated biorefineries, will be shaped by many of the same forces that shaped the development of the hydrocarbon economy and its refineries over the last century. The biggest difference between biofuels and petroleum fuels is the oxygen content. Biofuels have 10–45 wt% oxygen, whereas petroleum fuels have essentially none, making the chemical properties of biofuels very different than those from petroleum. Currently, there are two liquid biofuels in the market to partially replace petroleum: (1) ethanol can replace gasoline, and (2) biodiesel can replace diesel. Manufacturing processes for additional biofuels that are closer to petroleum will soon come into the market as described in the following chapters. Salient features of various biofuels are briefly described below.

1.5.1 Ethanol

Currently, ethanol is the most widely used liquid biofuel. It is produced by fermentation of sugars, which can be obtained from natural sugars (e.g., sugar cane, sugar beet), starches (e.g., corn, wheat), or cellulosic biomass (e.g., corn stover, straw, grass, wood). The most common feedstock is sugarcane or sugar beet, and the second common feedstock is cornstarch. Currently, cellulosic biomass use is very limited due to expensive pretreatment that is required for breaking the crystalline structure of cellulose. Bioethanol is already an established commodity due to its ongoing nonfuel uses in beverages and in the manufacture of pharmaceuticals and cosmetics. In fact, ethanol is the oldest synthetic organic chemical used by mankind. Table 1.12 shows ethanol production in different continents.

Table 1.12. *Global ethanol production in billion gallons in 2008 (Agranet, 2009)*

Country	Year 2008	Share year 2008 (%)
United States	8.93	43.8
Brazil	6.90	33.9
China	1.02	5.0
India	0.61	3.0
France	0.40	1.9
Canada	0.26	1.3
Germany	0.22	1.1
Thailand	0.15	0.7
Russia	0.15	0.7
Spain	0.13	0.6
South Africa	0.11	0.5
United Kingdom	0.11	0.5
Remaining countries	1.40	6.9
World	20.37	100.0

Ethanol can be used as a 10% blend with gasoline without need for any engine modification. However, with some engine modification, ethanol can be used at higher levels, for example, E85 (85% ethanol).

1.5.2 Methanol

Methanol is another conventional motor fuels. In the past, it has been considered a possible large-volume motor fuel substitute at various times during gasoline short-ages. It was often used in the early twentieth century to power automobiles before the introduction of inexpensive gasoline. Later, synthetically produced methanol was widely used as a motor fuel in Germany during World War II. Again, during the oil crisis of the 1970s, methanol blending with gasoline received attention due to methanol availability and low cost. Similar to ethanol, methanol has a high octane rating and hence is suitable for the Otto engine. Today, methanol is commonly used in biodiesel production for its reactivity with vegetable oils.

Before the 1920s, methanol was obtained from wood as a coproduct of char-coal production, and hence was commonly known as wood alcohol. But currently, methanol is manufactured worldwide from syngas (a mixture of hydrogen and car-bon monoxide gases derived from NG, refinery off-gas, coal, or petroleum) as

$$2H_2 + CO \rightarrow CH_3OH$$

The above reaction can be carried out in the presence of a variety of catalysts, including Ni, Cu/Zn, Cu/SiO$_2$, Pd/SiO$_2$, and Pd/ZnO. In the case of coal, it is first pulverized and cleaned, then fed to a gasifier bed where it is reacted with oxygen and steam to produce the syngas. A 2:1 mole ratio of hydrogen-to-carbon monoxide is fed to a fixed catalyst bed reactor for methanol production. Also, the technology for making methanol from NG is already in place and in wide use. Current NG feedstocks are so inexpensive that, even with tax incentives, renewable methanol has not been able to compete economically.

Renewable methanol can be produced directly from biomass or from biosyngas. Composition of biosyngas from biomass is shown in Table 1.13. The hydrogen-to-CO ratio in biosyngas is less than that from coal or natural gas, hence additional hydrogen is needed for full conversion to methanol. Part of gases produced from

Table 1.13. *Composition of biosyngas from biomass gasification*

Constituents	Volume % (dry and nitrogen-free basis)
Carbon monoxide (CO)	28–36
Hydrogen (H$_2$)	22–32
Carbon dioxide (CO$_2$)	21–30
Methane (CH$_4$)	8–11
Ethene (C$_2$H$_4$)	2–4

biomass can be steam reformed to produce hydrogen and followed by water–gas-shift reaction to further enhance hydrogen content.

Although methanol has a higher octane rating – Research Octane Number (RON) is 136 as compared with 129 for ethanol – it is more toxic and can be lethal, if ingested. Methanol vapors are poisonous, burn with invisible flame, and are five times more toxic than ethanol. The maximum allowable exposures to methanol and ethanol vapors are 200 ppm and 1000 ppm, respectively (Reed and Lerner, 1973). The water solubility of methanol becomes another environmental concern in case of a fuel spill. Methanol vapor is heavier than air and also flammable; hence, upon spills, the methanol vapor will tend to accumulate at ground level or in low-lying areas. Methanol is incompatible with many general materials normally used in petroleum storage and transfer systems, including aluminum, magnesium, rubber-ized components, and some gasket and sealing materials. Therefore, when used as an alternative fuel, it will be necessary to take special precautions to ensure that methanol is transported or stored in compatible materials designed for methanol. Energy density of methanol (16 MJ/liter) is only half of the gasoline energy density (32 MJ/liter), since methanol contains 50% oxygen (higher than biomass!).

1.5.3 Butanol

Butanol (C_4H_9OH) is a four-carbon alcohol compared to ethanol with two car-bon atoms. Due to the lower oxygen content, the energy density (29.2 MJ/liter) is much closer to that of gasoline. At present, butanol is presently manufactured from petroleum and primarily used as an industrial solvent. Historically (early 1900s–1950s), butanol was manufactured from corn and molasses by the well-known ABE (acetone, butanol, ethanol) fermentation process. However, as the price of petroleum dropped, production by fermentation declined. ABE fermentation by *Clostridium acetobutylicum* was the route used to produce butanol during World War II. In the current scenario, the historic ABE fermentation process for butanol is viewed as too expensive. One of the problems of butanol production from fer-mentation is low yield. The yield of butanol is only 1.3–1.9 gallons/bushel of corn, whereas yeast fermentation produces 2.7 gallons of ethanol/bushel of corn. Its low final concentration (1–2%) compares poorly with that of ethanol from yeast fermen-tation (10–15%). Butanol's boiling point (117°C) is higher even than that of water. At the 1–2% final batch concentration, there is a significant amount of water to boil off, which is expensive. With the current emphasis on renewable fuels, research interest has been rekindled recently in developing a viable ABE fermentation pro-cess (Huang, Ramey, and Yang, 2004).

Butanol is less miscible with water, less corrosive, and has lower heat of vapor-ization as compared with ethanol. The low aqueous solubility makes it compatible with petroleum infrastructures and could minimize the cosolvency concern associ-ated with ethanol, consequently decreasing the tendency of microbial-induced cor-rosion in fuel tanks and pipelines during transportation and storage. Viscosity is higher than that of ethanol and is comparable to that of high-quality diesel. Some of the disadvantages of butanol as fuels are the lower octane number (RON of 96) and the greater toxicity than ethanol. In case of spills, the danger of water contamination and its effect on the environment and human health need to be examined. That is why the use of butanol as fuel has yet to be approved by the various planning and regulatory agencies.

1.5.4 Biogas

Biogas is an environment friendly, clean, cheap, and versatile gaseous fuel. It is mainly a mixture of CH_4 and CO_2 obtained by anaerobic digestion of biomass, sewage sludge, animal wastes, and industrial effluents. Anaerobic digestion occurs in the absence of air and is typically carried out for a few weeks. Most of the biogas plants (Figure 1.8) use animal dung or sewage. Typical compositions of biogas and raw landfill gas are given in Table 1.14. CH_4 and CO_2 make up around 90% of the gas volume produced, both of which are greenhouse gases. However, CO_2 is recycled back by the plants, but CH_4 can significantly contribute to global warming. Hence, the capture and fuel use of biogas is beneficial in two ways: (1) fuel value, and (2) conversion of CH_4 into CO_2, which is a plant-recyclable carbon.

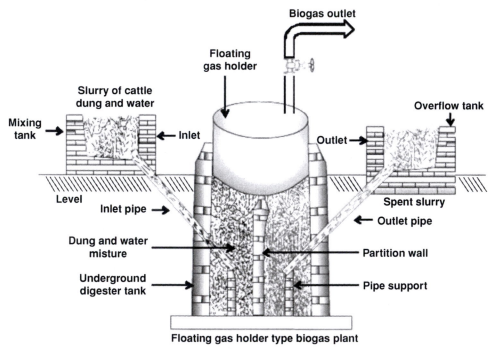

Figure 1.8. Biogas gas plant with floating gas holder (adapted from IGP, 2009).

1.5.5 Hydrogen

Hydrogen is not a primary fuel. It must be manufactured from water using energy from either fossil or nonfossil sources. Use of hydrogen fuel has the potential to

Table 1.14. *Composition of biogas and landfill gas (Monnet, 2003)*

Constituents	Unit	Biogas	Landfill gas
Methane (CH_4)	v%	55–70	45–58
Carbon dioxide (CO_2)	v%	30–45	32–45
Nitrogen (N_2)	v%	0–2	0–3
Volatile organic compounds (VOC)	v%	0	0.25–0.50
Hydrogen sulphide (H_2S)	ppm	~500	10–200
Ammonia (NH_3)	ppm	~100	0
Carbon monoxide (CO)	ppm	0	trace

improve both global climate and air quality. In addition, by using a fuel cell, hydrogen can be used with high energy efficiency, which is likely to increase its share as an automotive fuel (Gupta, 2008). A fuel cell is a device or an electrochemical engine that converts the energy of a fuel directly to electricity and heat without combustion

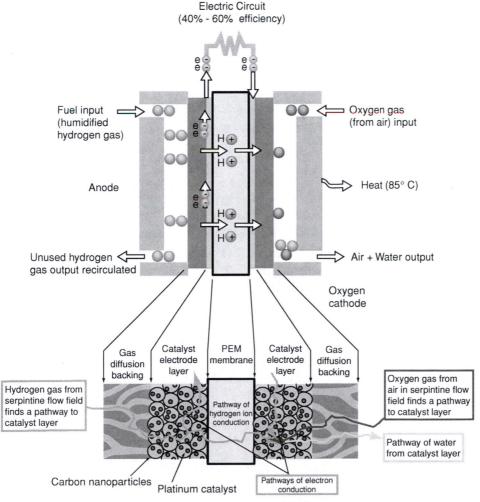

Figure 1.9. Schematic of fuel cell us hydrogen to produce electricity (reprinted from Jacobson, 2007).

(Figure 1.9). It consists of two electrodes sandwiched around an electrolyte. When hydrogen passes over one electrode and oxygen over the other, electricity is generated. The reaction product is water vapor and the fuel cells are clean, quiet, and efficient.

1.5.6 Biodiesel

Biodiesel is obtained from vegetable oils by reacting methanol using a process called transesterification. The purpose of the transesterification process is to lower the viscosity and oxygen content of the vegetable oil. In this process, an alcohol (e.g., methanol, ethanol, butanol) is reacted with the vegetable oil (fatty acid) in the presence of an alkali catalyst (e.g., KOH, NaOH) to produce biodiesel and glycerol. Being immiscible, biodiesel is easily separated from glycerol. Transesterification is an inexpensive way of transforming the large, branched molecular structure of the vegetable oils into smaller, straight-chain molecules of the type required in regular diesel combustion engines. Most of the biodiesel is currently made from soybean, rapeseed, and palm oils. The high value of soybean oil as a food product makes production of a cost-effective biodiesel very challenging. However, there are large amounts of low-cost oils and fats (e.g., restaurant waste, beef tallow, pork lard, and yellow grease) that could be converted to biodiesel. Biodiesel produces slightly lower power and torque; hence, the fuel consumption is higher than No. 2 diesel. However, biodiesel is better than diesel in terms of sulfur content, flash point, aromatic content, and biodegradability.

1.5.7 Bio-Oil

When powdered biomass is quickly heated (typically to 500°C), an oil-like substance comes out in the vapor phase. The substance, known as bio-oil, is obtained upon rapid cooling of the vapors. The oil has similar heating value as that of biomass, but has higher density than that of biomass. For use of biomass in a remote location, it is more economical to convert into bio-oil and then transport the bio-oil. Using the fast pyrolysis technology, a high yield of bio-oil can be achieved from various biomass feedstocks. However, this oil is unstable and acidic, contains char particles, and has about half the heating value of petroleum liquid fuels. Acidity, high viscosity, high oxygen content, difficulties in removing char particles, and immiscibility with petroleum liquids have restricted the use of this oil. Upon further upgrading of the bio-oil, it can be used directly in vehicle engines – either totally or partially in a blend.

1.5.8 Diesel from Fisher–Tropsch Technology

For use in transportation, liquid fuels are preferred over gaseous fuels due to the ease of handling and a high energy density. Upon gasification, various biomasses (irrespective of their differences) can be converted into a common biosynthesis gas.

The synthesis gas can be converted into liquid fuels using Fischer–Tropsch synthesis technology (FT). FT can be used to produce chemicals, gasoline, and diesel fuel. The FT products are predominantly linear molecules, hence the quality of the diesel fuel is very high. Because the purified synthesis gas is used in FT, all the products are free of sulfur and nitrogen. A major disadvantage of FT synthesis is the polymerization-like nature of the process, yielding in a wide product spectrum, ranging from compounds with low molecular mass, like methane, to products with very high molecular mass, like heavy waxes.

The first FT plants began operation in Germany in 1938 and operated until World War II. Since 1955, Sasol, a world leader in the commercial production of liquid fuels and chemicals from coal, has been operating FT processes. A major advantage of FT is the flexibility in feedstocks as the process starts with synthesis gas, which can be obtained from a variety of sources (e.g., NG, coal, biomass). Depending on the market conditions, a suitable feedstock can be selected and varied. In addition, FT can be tuned to produce a wide range of olefins, paraffins, and oxygenated products (e.g., alcohols, aldehydes, acids, ketones). For example, the high-temperature, fluidized-bed FT reactors with iron catalyst are ideal for the production of large amounts of linear olefins. FT started with one mole of CO reacting with two moles of H_2 to produce hydrocarbon chain extension ($-CH_2-$), as follows:

$$CO + 2H_2 \rightarrow -CH_2- + H_2O \quad \Delta H = -65 \, kJ/mol$$

The $-CH_2-$ is a building block for longer hydrocarbons. The overall reactions can be written as

$$nCO + 2nH_2 \rightarrow [-CH_2-]_n + nH_2O \quad nCO + (2n+1)H_2 \rightarrow C_nH_{2n+2} + nH_2O$$
$$nCO + (n+m/2)H_2 \rightarrow C_nH_m + nH_2O$$

From the above exothermic reactions, a mixture of paraffins and olefins is obtained. Typical operating temperature range is 200–350°C and pressure range is 15–40 bar. Iron catalysts have a higher tolerance for sulfur, are cheaper, and produce more olefin and alcohols. However, the lifetime of the iron catalysts is short (e.g., typically 8 weeks in commercial installations).

1.5.9 Biocrude

Biomass can be depolymerized using hydrothermal treatment into its monomer constituents. Upon hydrothermal treatment, aromatic monomers from lignin, glucose monomers from cellulose, and xylose monomers from hemicelluloses are obtained. The liquid, termed biocrude, contains 10–20 wt% oxygen and 30–36 MJ/kg heating value as opposed to <1 wt% and 42–46 MJ/kg for petroleum. The high oxygen content imparts lower energy content and poor stability. Hence, biocrude needs to be deoxygenated to make it compatible with conventional petroleum. Because biocrude contains less oxygen and more heating value than biomass, the hydrothermal liquefaction process is also referred to as hydrothermal upgrading process. As compared with bio-oil from the fast pyrolysis, biocrude produced from hydrothermal liquefaction has higher energy value and lower moisture content, but requires longer residence time and higher capital costs. Typical hydrothermal lique-faction conditions range from 280 to 380°C, 70 to 200 bar with liquid water present, and reaction occurring for 10–60 minutes.

1.5.10 Biochar

Heating (to about 300–500°C) biomass in the absence of oxygen (pyrolysis) can convert biomass into biochar, a charcoal material. As compared with 40% carbon in biomass, biochar can contain up to 80% carbon. With reduced oxygen content, heating value of biochar is significantly higher than that of biomass (as much as 1.7-fold). Biochar can be used as a carbon-neutral feedstock for power generation in electricity plants. In addition, biochar can be added to agricultural soil to increase the fertility, thereby reducing the use of fossil fuel-based fertilizer.

1.6 Summary

Currently, much of our primary energy need is being supplied by fossil sources. Unfortunately, the heavy use of fossil fuels has given rise to atmospheric pollution and greenhouse gas emissions. In addition, the present petroleum reserves will last only up to 50 years at the current rate of consumption. Hence, there is an urgency to develop use of the renewable energy resources out of which biomass represents an important alternative. Various technologies exist through which biomass can be converted into most preferred liquid form of the fuel, including ethanol, diesel, and gasoline. The share of renewable energy sources is expected to increase significantly in coming decades.

2 Air Pollution and Global Warming from the Use of Fossil Fuels

2.1 Introduction

Today's world is facing two environmental problems: global warming and air pollution. Both of these are linked to the large-scale use of fossil fuels. Energy use is being reevaluated because of the quadrupling of oil prices in the 1970s and rapid increase in 2008, the growing awareness of energy-related pollution, and the possibility of climate change. As a result, there has been a significant improvement in energy efficiency in all sectors, including industry, power generation, lighting, household appliances, transportation, and climate control of buildings. In fact, efficient use of energy has historically helped increase the energy intensity in all Organization for Economic Cooperation and Development (OECD) countries and more recently in transition economies. Unfortunately, a rapid increase in the per capita use of fossil energy has accelerated the rate of atmospheric pollution.

2.2 Air Pollution

Air pollution (Figure 2.1) is due to the harmful substances (natural or man-made) that we breathe, including fine particles (produced from burning of fossil fuels), ground-level ozone (a reactive form of oxygen that is a component of urban smog), and gases such as carbon monoxide, chemical vapors, nitrogen oxides, and sulfur dioxide. The ill effects of air pollutants on human health have been known for the last 30 years; these include asthma, cardiovascular diseases, decreased lung function, and allergies. Of urgent concern is the long-term ill effect on lung development in children. In addition, air pollution causes damage to plant and animal lives.

For protection, the Air Quality Standard (AQS) has been developed, which prescribes the safe level of pollutants in the outside air. Any concentration of pollutant gases and suspended particles in excess of AQS is considered harmful. Guidelines for the pollutant concentration are provided in Table 2.1 as suggested by the World Health Organization (WHO).

Table 2.1. *Air quality guidelines for particulate matter, ozone, nitrogen dioxide, and sulfur dioxide (WHO, 2005)*

Pollutant	Permissible limit	Duration
Particulate matter PM_{10}	$10 \ \mu g/m^3$	Annual mean
(thoracic fraction, diameter $\leq 10 \ \mu m$)	$25 \ \mu g/m^3$	24-h mean
Particulate matter $PM_{2.5}$	$20 \ \mu g/m^3$	Annual mean
(respirable fraction, diameter $\leq 2.5 \ \mu m$)	$50 \ \mu g/m^3$	24-h mean
Ozone (O_3)	$100 \ \mu g/m^3$	8-h mean
Nitrogen oxides	$40 \ \mu g/m^3$	Annual mean
	$200 \ \mu g/m^3$	1-h mean
Sulfur dioxide	$20 \ \mu g/m^3$	24-h mean
	$500 \ \mu g/m^3$	10-min mean

Sulfur dioxide pollutant comes from the combustion of fossil fuels, as coal and petroleum contain various sulfur compounds, including benzothiophenes, dibenzothiophenes, benzonaphthothiophenes, phenanthrothiophenes, and other sulfur-containing aromatic molecules. The following are the key pollutants from the combustion of coal:

• Nitrogen oxides
• Sulfur oxides
• Carbon dioxide
• Fine particulates
• Heavy metals (mercury, lead, etc.)

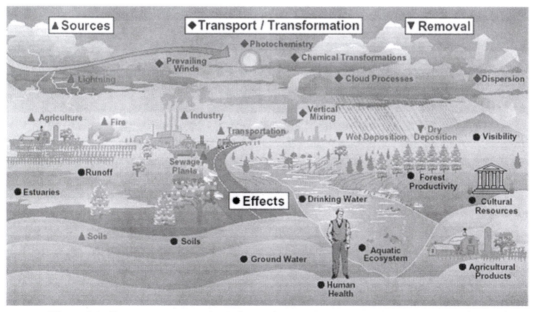

Figure 2.1. Sources, transport/transformation, and removal of air pollutants (U.S. EPA, 2009a).

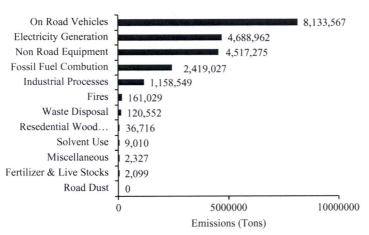

Figure 2.2. Tons of NO$_x$ emitted in the United States in 2002, from various sources (U.S. EPA, 2009b).

2.2.1 Nitrogen Oxides

When exposed to high temperatures (e.g., in a combustion chamber), some nitrogen in the air oxidizes and forms NO, NO$_2$, and N$_2$O gases. NO$_x$ represents nitric oxide (NO) and nitrogen dioxide (NO$_2$), which are responsible for the formation of acid rain, photochemical smog, and ground-level ozone. NO$_x$ is formed at elevated temperatures ($\sim$1,200°C) as found in pulverized coal flames, whereas nitrous oxide (N$_2$O) is formed at lower temperatures ($\sim$800°C) like those found in fluidized bed combustors. The atomic nitrogen present in the fuel is oxidized at all combustion temperatures (Kutz, 2007). Figure 2.2 shows the emissions of total NO$_x$ in the United Sates in 2002 (UNEP, 2007).

2.2.2 Sulfur Dioxide

When petroleum (especially diesel or heating oil) and coal are burned, the sulfur contained also burns to sulfur dioxide (SO$_2$). Coal combustion makes up the majority of SO$_2$ emissions from utilities, which in turn account for about one-third of the total SO$_2$ emissions from all sources (Figure 2.3). In addition to the organically bonded sulfur, coal can also contain pyritic sulfur.

2.2.3 Fine Particles

Out of the non-natural process of fine particle pollution, coal-burning plants (especially those used for power production) and automobiles produce significant amounts of fine particulate matters (PMs). The particles arise from the ash content in the coal and from the unburned carbon. The particles that are smaller than 1 μm are the most hazardous because these can penetrate into the alveolar region of the lungs. Particles smaller than 2.5 μm can penetrate into the lungs, and particles smaller than 10 μm can settle in the bronchial tubes of the lungs. Fine particles

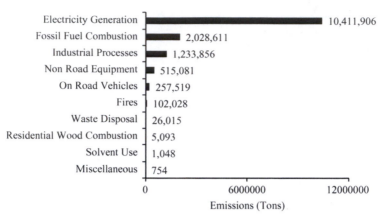

Figure 2.3. Tons of SO_2 emitted in the United States in 2002, from various sources (U.S. EPA, 2009c).

can cause significant health damage if exposure occurs over an extended period. Figure 2.4 shows concentration of particulate matter less than 2.5 μm ($PM_{2.5}$) for different Asian cities for year 2001–2004. The WHO limit for normal ambient air for $PM_{2.5}$ exposure is 20 μg/m^3 averaged annually (Table 2.1). In all six cities, the levels were found to be high, especially during the dry season, and frequently exceeded the WHO standards at a number of sites. The average concentrations of $PM_{2.5}$ ranged from 44 to 168 in the dry season and from 18 to 104 μg/m^3 in the wet season (Kim Oanh et al., 2006).

2.2.4 Mercury

From the exhaust of coal combustion, most heavy metals, including lead, tin, and magnesium, are collected in particulate control systems, but mercury escapes into the environment due to its low vaporization temperature (356°C), making coal combustion the primary source for mercury emissions. The worldwide average mercury content in coal is 0.10 ppm (Yudovich and Ketris, 2005). However, there are coals that contain 10- to 100-fold more mercury, for example, those located in Donbas (Ukraine), the Appalachian basin and Texas (U.S.), the Russian Far East, and Southern China. Mercury from the flue gas precipitates out into the environment and bioaccumulates in organisms such as fish, where it is often transformed into methylmercury, a highly toxic organic compound. Methylmercury bioaccumulates into the internal organs, including the liver, and into muscle tissues. Fish species that are high up on the food chain contain high concentrations of mercury because they eat many smaller fish that have small amounts of mercury in them (Figure 2.5). The permissible limit for mercury content in fish suitable for food is 1 ppm in the United States and 0.5 ppm in Canada (Kutz, 2007). In humans, methylmercury exposure can cause permanent neurological damage, leading to loss of coordination, reproductive disorders, paralysis, cerebral palsy, and impaired hearing, speech, and vision. Severe exposure can lead to coma or even death.

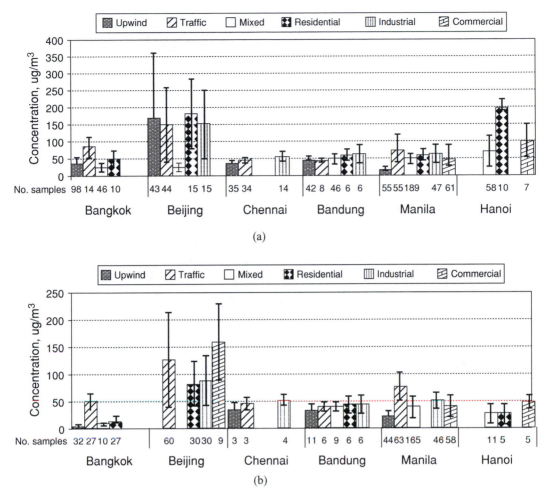

Figure 2.4. Average concentrations of PM$_{2.5}$ in dry (a) and wet (b) seasons in six Asian cities during 2001–2004. Whiskers represent one standard deviation (SD) of the mass concentration data series at each site. Note that SD mainly indicates the variations in daily concentrations due to changing emission and meteorology conditions throughout a season (Kim Oanh et al., 2006; reproduced with permission from Elsevier).

2.2.5 Lead

Sources of lead pollution are leaded petrol and diesel, paints, lead batteries, hair dye, and so on. However, many countries have now prohibited the use of lead in petroleum or paints. In some parts of South America, Asia, Eastern Europe, and the Middle East, leaded gasoline is still in use. Addition of tetraethyl lead has been used to improve the octane rating of petroleum fuels. In addition, lead provides valve lubrication. Compared with most other pollutants, lead remains in the atmosphere for a long time after release, due to the lack of easy degradation pathways. Hence, lead remains available to impact the food chain and human metabolism for an extended period (Sauve et al., 1997). The permissible limit for lead in the respirable air is 1 μg/m^3. The toxic effects of lead include damage to the kidney and

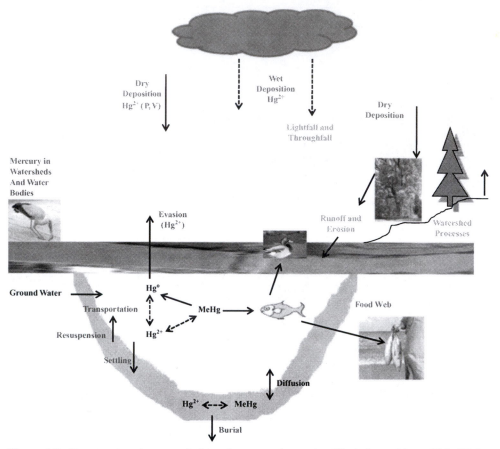

Figure 2.5. Transport and accumulation of mercury in marine life (adapted from U.S. EPA, 2009d).

cardiovascular and nervous systems. Of particular concern is the effect on cognitive and behavioral development in children, even at a relatively low exposure (Pirkle et al., 1998).

2.3 Carbon Dioxide Emissions

About 98% of the global carbon dioxide (CO_2) emissions result from fossil fuel (coal, oil, and natural gas) combustion. A typical hydrocarbon combustion reaction can be written as

$$C_xH_y + (x + 0.25y)O_2 \rightarrow xCO_2 + 0.5yH_2O + \Delta H_c \qquad (2.1)$$

where the heat of combustion (ΔH_c) depends on the carbon:hydrogen (x:y) ratio in the fuel. A fuel containing less hydrogen produces more CO_2 to obtain the energy. To compare CO_2 emissions from various fuels on an equal energy basis, CO_2 emission factors (CEF) are calculated. In most cases, CEFs are defined as the mass of carbon (C) in the fuel per energy unit, for example, kg C/GJ. To calculate the emission as CO_2 instead of C, the emission factor is multiplied by the ratio of the

Table 2.2. *Top ten CO_2-emitting countries in 2006 (CDIAC, 2009)*

Country	Total emissions (million metric ton CO_2)	Per capita emission (tons CO_2/capita)
China	6,101	4.7
United States	5,749	19.0
Russia	1,565	11.0
India	1,510	1.4
Japan	1,293	10.3
Germany	802	9.8
Canada	546	16.7
United Kingdom	568	9.4
South Korea	476	9.8
Iran	465	6.6

molecular weight of CO_2 to the atomic weight of carbon (44:12). Typical CEFs in C/GJ units are 15 for natural gas, 20 for crude oil, 25 for hard coal, and up to 30 for lignite (Hiete et al., 2001). Hence, natural gas is the cleanest and lignite is the dirtiest fossil fuel in terms of CO_2 emission. At the present time, coal is responsible for 30–40% of world CO_2 emissions from fossil fuels. Figure 2.6 shows the global emissions of CO_2 from fossil fuels. Recently, the Asia-Pacific region has shown the highest rate of increase due to development in India and China. However, per capita CO_2 emissions in Asia are very low compared with those in North America or Europe (Table 2.2).

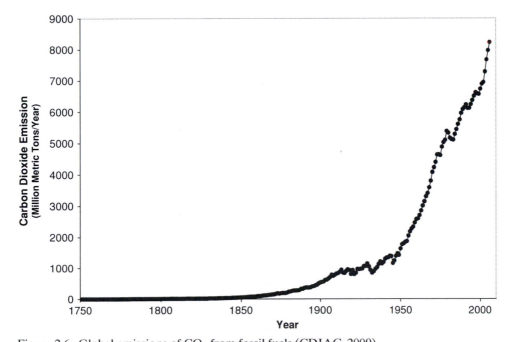

Figure 2.6. Global emissions of CO_2 from fossil fuels (CDIAC, 2009).

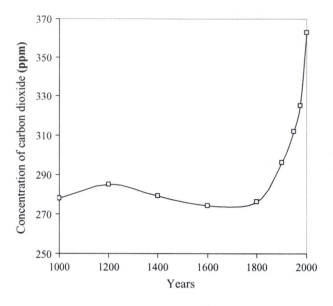

Figure 2.7. World CO_2 concentrations between years 1000 and 2000 (adapted from Jean-Baptiste and Ducroux, 2003).

Emitted CO_2 has accumulated in the earth's air, causing the CO_2 concentration to increase. The increase over the last one thousand years is shown in Figure 2.7, and Figure 2.8 shows the increase since 1959.

As discussed in the following text, the increase in CO_2 concentration has been a cause of significant concern due to its contribution to global warming. A

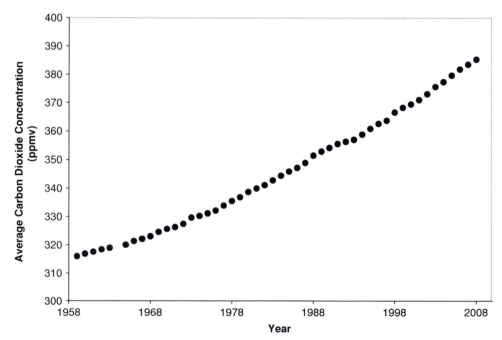

Figure 2.8. Annual average CO_2 concentration measured at Mauna Loa (Hawaii, U.S.) (CDIAC, 2009).

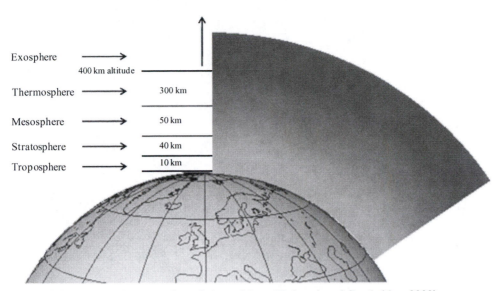

Figure 2.9. The earth's atmosphere (adapted from University of Cambridge, 2009).

reduction in the use of fossil fuels would considerably reduce the emission of CO_2 and other pollutants. This reduction can be realized by increasing the efficient use of fossil energy and/or by using carbon emission–free energy sources, including nuclear, solar, wind, geothermal, and biomass energy. Although biomass combustion does involve the release of CO_2, the equivalent amount of CO_2 is then removed from the air by growing plants.

2.4 Greenhouse Effect

The earth receives energy from the sun in the form of solar radiation. If the earth did not have atmosphere, the areas receiving solar radiation would be extremely hot (during the daytime) and those not receiving radiation would be extremely cold (during the night), similar to those on the moon. The earth's atmosphere can be divided into several layers (Figure 2.9), out of which, the troposphere sustains life.

The earth has a natural temperature control due to the presence of atmosphere, which contains greenhouse gases (GHGs). Earth absorbs about two-thirds of the solar radiation and reflects the remaining. Earth's warm surface radiates energy as infrared radiation (Figure 2.10). GHGs absorb the infrared radiation, thus warming the atmosphere. Naturally occurring GHGs are CO_2, water vapor, ozone, methane (CH_4), SO_2, and N_2O. The GHG molecules contain three or more atoms and have a higher heat capacity than oxygen (O_2) or nitrogen (N_2) molecules. For example, the ambient heat capacities of CO_2, CH_4, and SO_2 are 3.40, 3.34, and 3.76 cal/°C mol, respectively, which are much higher than those of N_2 or O_2 (2.49 and 2.51 cal/°C mol, respectively). Once released into the atmosphere, GHGs can persist for a long time, depending on their molecular vulnerability. Table 2.3 shows the concentration, source, and global warming potential of various GHGs.

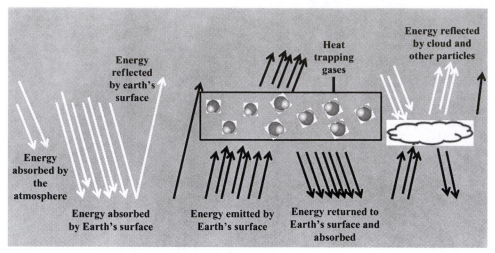

Figure 2.10. The greenhouse effect (adapted from UCS, 2009).

Historically, CO_2 has been the most prominent GHG, but CH_4, N_2O, and chlorofluorocarbons (CFCs) have assumed increasing significance. Due to the concern for the ozone layer depletion, hydrofluorocarbons (HFCs) have been widely approved as a substitute for CFCs. Nonetheless, trace gases including HFCs, perfluorocarbons (PFCs), and sulfur hexafluoride (SF_6) are to be regulated under the 1997 Kyoto Protocol due to the potential for an increase in atmospheric concentration because of their long half-lives.

Due to human activities, atmospheric CO_2 concentration has increased by more than 33% over the last 100 years. And now a large number of scientists attribute global warming to this increase in CO_2 concentration. On the other hand, some people believe that proof for the global warming theory is incomplete due to uncertainties in the natural variation in the earth's climate. For example, climate is affected by the way heat and energy distribute among the land, ocean, atmosphere, and ice components of the earth. In addition, one needs to look at the dynamic distribution of carbon on the earth. Figure 2.11 shows the present global carbon cycle.

The global carbon cycle shows the carbon reservoirs in GtC (gigaton = one billion tons) and fluxes in GtC/year. These values can be converted into equivalent CO_2 concentration by multiplying by the 44:12 factor. The values shown in Figure 2.11 are annual averages during the period 1980–1989. The evidence shows that many of the fluxes can fluctuate significantly from one year to another. In fact, the carbon system is dynamic and is coupled to the climate system on seasonal, interannual, and decadal time scales.

To reduce CO_2 emission for environmental protection, there is an urgent need to find alternatives to fossil fuel. Because it seamlessly fits into the current carbon-based energy use, biomass appears to be an obvious choice. Scientists believe that using biomass in a sustainable manner will not cause any net increase in atmospheric CO_2, as the CO_2 given off by biomass combustion will be removed from the atmosphere by photosynthesis to produce the next crop of biomass. In fact, some even

Table 2.3. *Main greenhouse gases (IPCC, 1996)*

Greenhouse gas	Concentration (ppb[a] by volume)		Atmospheric lifetime (years)	Anthropogenic source	Global warming potential	Contribution to global warming
	Before 1750	In 2007				
Carbon dioxide (CO_2)	278,000	384,000	100	Fossil fuel combustion; land use conversion; cement production	1	55%
Methane (CH_4)	700	1,857	12	Fossil fuel; rice paddles; waste dumps; livestock	21	15%
Nitrous oxide (N_2O)	270	321	114	Fertilizer; industrial processes; combustion	310	6%
CFC-12 (CCl_2F_2)	0	0.541	102	Liquid coolants; foams	6,200–7,100	24%
HCFC-22 ($CHClF_2$)	0	0.197	12	Liquid coolants	1,300–1,400	
Perfluoromethane (CF_4)	0	0.080	50,000	Production of aluminum	6,500	
Sulfur hexafluoride (SF_6)	0	0.064	3,200	Dielectric fluid	23,900	

[a] ppb, parts per billion.

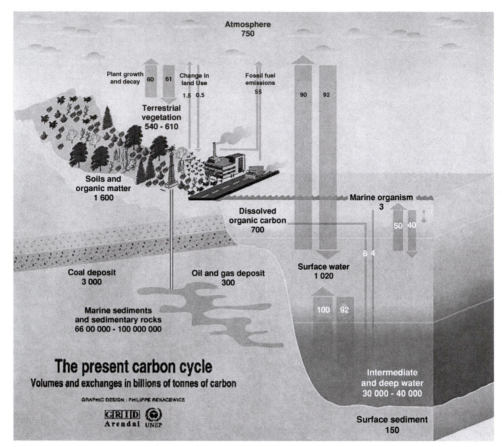

Figure 2.11. The present global carbon cycle (IPCC, 1996).

believe that sustainable use of biomass will result in a net decrease in atmospheric CO_2 by sequestering carbon in the soil upon root growth.

2.5 Global Warming

Based on measurement over the last century, average global temperature has increased by 0.56°C. This increase is referred to as global climate change or global warming. Figure 2.12 shows the change in the global land and ocean surface temperatures relative to the period 1951–1980. Due to an increase in CO_2 emission and an auto-feedback mechanism of heating, the increase in the temperature for the next century is estimated to be anywhere from 1.5°C to 5.8°C. Such a drastic warming can result in the following:

- Displacement of agricultural zones
- Migration of tropical disease areas
- Melting of polar ice caps and glaciers
- Rise in sea level of 9–88 cm

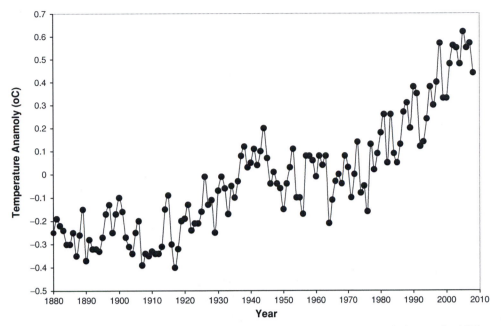

Figure 2.12. Change in the global land and ocean surface temperatures relative to the 1951–1980 period (Goddard Institute for Space Studies, 2009).

The increase in GHGs can be attributed to production and human use of energy (60%), use of chemicals such as CFCs (15%), agriculture (12%), land-use modifications (9%), and other human activities (4%).

If the sea level rises as predicted, large metropolitan centers, such as New York, Los Angeles, London, Mumbai, and Tokyo, may eventually be covered with water. With the increase in the world population and the overall standard of living, energy consumption is also likely to increase dramatically. Increasing energy demand is being met by the use of more coal. If the coal industry is to participate in this growth, it is necessary to address the issue of CO_2 disposal.

The relationship of temperature and CO_2 concentration (data obtained from the ice core) in the atmosphere over the last 400,000 years is shown in Figure 2.13. As seen in these data, the earth's climate has seen significant fluctuations with very significant temperature variations, going from a warm climate to an ice age in a few decades. These rapid changes indicate that the climate may be quite sensitive to internal and/or external forces and feedbacks. It is interesting to note that the temperature has been less variable during the last 10,000 years, with a variation of less than 1°C percentury. The graph presents a strong correlation between atmospheric CO_2 concentration and temperature. A possible scenario can be deduced that anthropogenic emissions of GHGs could bring the climate to a state where it reverts back to the highly unstable climate of the pre–ice age period. Climate change has followed a highly nonlinear path in which sudden and dramatic surprises occur when GHG levels reach a new trigger point.

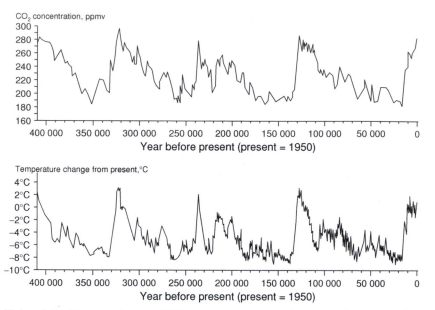

Figure 2.13. Temperature and CO_2 concentration in the astrosphere over the last 400,000 years (Petit et al., 1999; reproduced with permission).

2.5.1 Argument against Global Warming

There is a very small group of people who believe that global warming is not occurring. Their arguments include the following: (1) Humans cannot possibly influence the climate of such a vast earth that has seen millions of dramatic changes. (2) The present temperature increase is part of a larger natural cycle; GHG emissions from nonhuman activities (e.g., volcano eruptions) are much larger and uncontrollable. (3) According to the Big Bang theory, the earth is cooling. (4) Much of the universe is extremely cold (average temperature is about $-269°C$); hence, it is important to have more heat on the earth. (5) The GHG effect of water dissolved in the air and clouds is highly variable and most influential. (6) The climatic modeling has significant errors due to the complexity of global climate. This debate is further supported by special interest groups who would be negatively affected by possible legislation limiting CO_2 emission. Nonetheless, there is no conflicting view on the observation that human activities have caused a significant increase in the atmospheric CO_2 concentration. Some people argue that the increased CO_2 concentration will help us grow more food needed for the increasing population.

2.6 Kyoto Protocol

The global climate change issue is a subject of intense international debate and political activity, as the climatic consequences will be felt by all countries; the effect will not be limited to only those who emitted CO_2. At the World Summit in December 1997, attended by 160 countries, a resolution known as the Kyoto Protocol was passed to address the issue of GHG emission. According to the Kyoto Protocol, OECD member countries and those with transition economies must reduce their emissions of six GHGs by at least 5.2% compared with 1990 levels. The emission reduction should take place from 2008 to 2012. On the other hand, developing economies are not required to make any reduction commitments, as the per capita emissions in those economies are much lower than in the developed economies. The Kyoto Mechanism foresees the following:

- Domestic reductions in industrialized countries
- Flexibility mechanisms
- Exchanges among industrialized countries
- Trading, based on GHG emissions quota
- Joint implementation, based on projects to mitigate GHG emissions
- Exchanges between industrialized and developing countries
- A clean development mechanism, based on projects to mitigate GHG emissions
- Incentives and regulations and public–private partnerships
- Increased research and development of new technologies
- Energy efficiency and renewable energy
- GHG emission and energy-use taxes

2.7 Carbon Credits

Based on the assessment of a country's carbon emission and its required reduction, authorities can issue a carbon emission quota to various GHG-emitting units (i.e., companies, organizations). GHG emitters can reduce the emissions on their own by using renewable fuel and/or process efficiencies. Additionally, the GHG emitters can buy carbon credits from other companies who are able to reduce GHG emissions much more than required. In fact, businesses are emerging that can specialize in producing the credits, for example, by participating in reforestation or replacement of coal-based electricity by that produced from biomass, wind, or solar sources. The carbon credits are traded on "exchanges." The trading of carbon credits allows efficient reduction in GHG emissions, as the companies with reduction expertise can do the work for those who do not have such expertise. The use of biomass for energy presents a viable solution to meet the protocol requirements. If biomass could be converted into useful energy, the consumption of fossil fuel and GHG emissions would be decreased (Saga et al., 2008).

2.8 Carbon Sequestration

Because almost all aspects of modern life are linked to the use of fossil fuel, it will take a long time to reduce the use of fossil fuels. Unfortunately, at the same time, CO_2 emission will continue to take place, adding to the CO_2 concentration in the atmosphere. Hence, to stabilize and ultimately reduce the CO_2 concentration in the air, it will be necessary to sequester (permanently store) the produced CO_2 rather

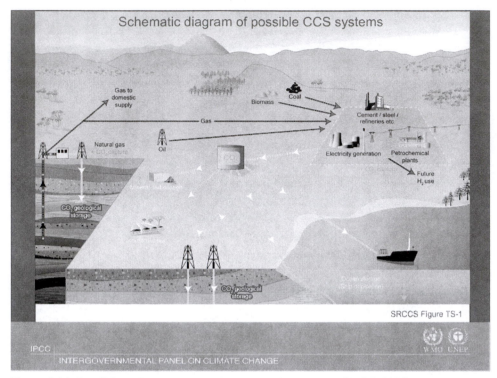

Figure 2.14. Possible carbon capture and sequestration systems (IPCC, 2005).

than releasing it into the air. Carbon sequestration, along use of fuels with reduced carbon content and improved efficiency of energy production and use, must play a major role if the nation is to enjoy the economic and energy security benefits that fossil fuels bring to the energy mix. The following are key elements of CO_2 sequestration: (1) cost-effective CO_2 capture and separation processes, (2) CO_2 sequestration in geological formations, including oil and gas reservoirs, (3) direct injection of CO_2 into the deep sea and stimulation of phytoplankton growth, (4) improved full life-cycle carbon uptake of terrestrial ecosystems, (5) advanced chemical, biological, and decarbonization concepts, and (6) models and assessments of the cost, risks, and potential of carbon sequestration technologies. A schematic of possible carbon capture and sequestration is shown in Figure 2.14.

2.9 Summary

The use of fossil fuel has given rise to the pollution of land, water, and air. The presence of these pollutants has adversely impacted human health and the environment. Emissions of GHGs, such as CO_2, are threatening the climatic stability via global warming. Global CO_2 concentration has been steadily increasing since industrialization (i.e., the start of fossil fuel use). To avoid dramatic changes in the climate, large reductions in CO_2 emissions are needed. Mechanisms are currently under development to limit CO_2 emissions.

3 Renewable Energy Sources

3.1 Introduction

Although fossil fuels (petroleum, coal, and natural gas) are still being created at a geological rate, they are being consumed at a much faster rate. Hence, fossil fuels are not renewed fast enough to be replenished. That is why fossil fuels are considered nonrenewable. Presently on Earth, renewable sources of energy are due to the sun's energy and its direct and indirect effects on the earth (solar radiation, wind, falling water, and various plants, i.e. biomass), gravitational forces (tides), and the heat of the earth's core (geothermal). Renewable energy sources (RES) will replenish themselves within our lifetime and may be used with suitable technology to produce predictable quantities of energy when required. Additionally, renewable resources are more evenly distributed than fossil and nuclear resources on the earth. The most important benefit of renewable energy systems is the decrease of environmental pollution.

The main RES and their usage forms are shown in Table 3.1. Some of these energy sources have been used by humans for over five thousand years. During the preindustrial era, RES were primarily used for heating and simple mechanical applications that did not reach high energy efficiency. In the industrial era, the energy use shifted from low energy value RES to much higher energy value coal and petroleum. In fact it is said that RES such as water and wind power probably would not have provided the same fast increase in industrial productivity as fossil fuels (Edinger and Kaul, 2000). Biomass now represents only 3% of the primary energy consumption in industrialized countries. However, much of the rural population in developing countries, which represents about 50% of the world's population, is reliant on biomass, mainly in the form of wood and stalk, for fuel (Ramage and Scurlock, 1996).

With the increase in environmental pollution due to the increased use of fossil energy, RES are emerging as promising alternatives. Current use of RES and their projections until 2040 are shown in Table 3.2. Currently, RES account for 13.6% of global energy use, and by 2040, this contribution is expected to be 47.7%. The most significant developments in renewable energy production are expected in

Table 3.1. *Main renewable energy sources and their usage forms*

Energy source	Energy conversion and usage options
Hydropower	Power generation
Modern biomass	Heat and power generation, pyrolysis, gasification, anaerobic digestion
Solar	Solar home system, solar dryers, solar cookers
Direct solar	Photovoltaics, thermal power generation, water heaters
Tidal	Barrage, tidal stream
Wind	Power generation, wind generators, windmills, water pumps
Wave	Numerous designs
Geothermal	Urban heating, power generation, hydrothermal, hot dry rock

photovoltaics [from 0.2 to 784 million ton oil equivalent (Mtoe)] and wind energy (from 4.7 to 688 Mtoe) between 2001 and 2040.

3.2 Biomass

During the process of photosynthesis, plants absorb atmospheric carbon dioxide (CO_2) and ground water to produce oxygen (O_2) and biomass. Dry biomass mostly contains carbon, hydrogen, and oxygen in the ratio 1:1.4:0.6 with overall chemical formula $[CH_{1.4}O_{0.6}]_n$ as a carbohydrate polymer. The fuel and feed use of biomass results in the release of CO_2 back into the atmosphere. CO_2 is then reused by the growth of the next crop of biomass in a cyclical manner (Figure 3.1). Because the annual average concentration of CO_2 in the atmosphere theoretically remains constant in this cycle, biomass energy is expected to become one of the key sources of sustainable energy in the future. That is why, in this decade, interest in the production of biomass and its use as energy has increased. Additionally, replacing fossil fuel with biomass would avoid the increase of CO_2 in the atmosphere and would help meet the obligations of the Kyoto Protocol.

Table 3.2. *Global renewable energy scenario by 2040 (Mtoe) (EREC, 2006)*

Renewable energy resource	2001	2010	2020	2030	2040
Biomass	1,080	1,313	1,791	2,483	3,271
Large hydro	22.7	266	309	341	358
Geothermal	43.2	86	186	333	493
Small hydro	9.5	19	49	106	189
Wind	4.7	44	266	542	688
Solar thermal	4.1	15	66	244	480
Photovoltaic	0.2	2	24	221	784
Solar thermal electricity	0.1	0.4	3	16	68
Marine (tidal/wave/ocean)	0.05	0.1	0.4	3	20
Total renewable energy sources	1,365.5	1,745.5	2,694.4	4,289	6,351
Total nonrenewable energy sources	8,672.5	8,803.5	8,730.6	8,063	6,959
Total energy sources	10,038	10,549	11,425	12,352	13,310
Renewable energy sources contribution (%)	13.6	16.6	23.6	34.7	47.7

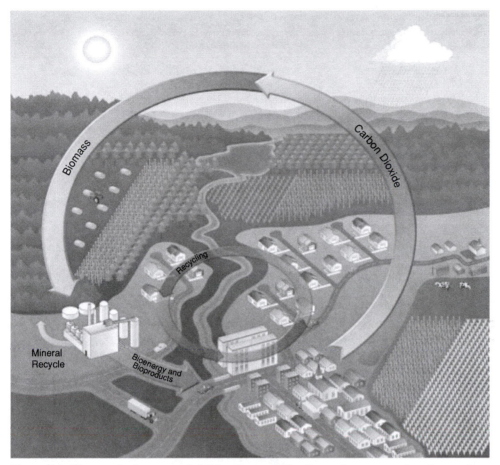

Figure 3.1. Biomass in forests and fields takes up CO_2. It can be converted to low-carbon fuels, construction wood, and other products (ORNL, 2000). Illustration by Brett Hopwood.

Typically, biomass refers to the nonfood part of plants. Various biomass resources include woody and herbaceous species, wood wastes, agricultural and industrial residues, waste paper, municipal solid waste, biosolids, waste from food processing, animal wastes, aquatic plants and algae, and so on. Major organic components of biomass can be classified as cellulose, hemicelluloses, and lignin. The major categories of biomass feedstock are listed in Table 3.3. The use of biomass energy can be placed in four major categories: wood and wood wastes (64%), municipal solid waste (24%), agricultural waste (5%), and landfill gases (5%) (Demirbas, 2000c).

World production of biomass is estimated at 146 billion metric tons per year, mostly in the form of wild plant growth. However, there is a significant potential for producing biomass from farm crops, as some farm crops and trees can produce up to 20 metric tons of biomass per acre each year (Cuff and Young, 1980). According to the World Energy Council, in 2001, global traditional biomass reserves were 930 Mtoe (WEC, 2004). A detailed description of global biomass availability is given in Chapter 4. The geographic availability of biomass in the United States is shown in Figure 3.2.

Table 3.3. *Major categories of biomass feedstock*

No.	Major category	Biomass feedstock
1	Forest products	Wood, logging residues, trees, shrubs and wood residues, sawdust, bark, etc.
2	Biorenewable wastes	Agricultural wastes, mill wood wastes, urban wood wastes, urban organic wastes
3	Energy crops	Short-rotation woody crops, herbaceous woody crops, grasses, forage crops
4	Food crops	Residue from grains and oil crops
5	Sugar crops	Sugar cane, sugar beets, molasses, sorghum
6	Landfill	Municipal solid wastes
7	Industrial organic wastes	Plastic wastes, oil wastes, leather wastes, rubber wastes, organic acid wastes, etc.
8	Algae, kelps, lichens, and mosses	Water hyacinth, mushrooms, etc.
9	Aquatic plants	Algae, water weed, water hyacinth, reed, and rushes

There are four major ways to use biomass energy: direct combustion, gasification, pyrolysis, and anaerobic digestion. Table 3.4 summarizes biomass technology options and corresponding end-uses, together with an indication of the status of these technologies. In countryside regions of developing nations, combustion of dry

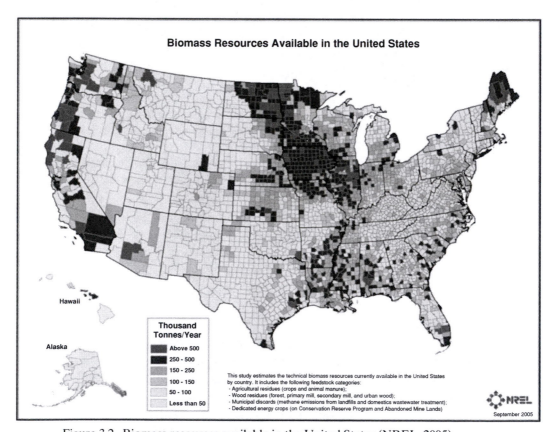

Figure 3.2. Biomass resources available in the United States (NREL, 2005).

Table 3.4. *Biomass technology options, corresponding end-uses and status (Bauen et al., 2004)*

Conversion technology	Resource type	Examples of fuels	Product	End-use	Technology status
Combustion	Mainly dry biomass	Wood logs, chips and pellets, agricultural residues, chicken litter	Heat	Heat, electricity (steam turbine, reciprocating steam engine, Stirling engine)	Commercial
Anaerobic digestion	Wet biomass	Manure, sewage sludge	Biogas and by-products	Heat (boiler), electricity (engine, gas turbine, fuel cell), transport fuel	Commercial
Gasification	Mainly dry biomass	Wood chips and pellets, agricultural residues	Product gas	Heat (boiler), electricity (engine, gas turbine, combined cycles, fuel cell), transport fuels (methanol, hydrogen)	Demonstration/ early commercial
Pyrolysis	Mainly dry biomass	Wood chips and pellets, agricultural residues	Pyrolysis oil, product gas, and char	Heat (boiler), electricity (engine, turbine)	Demonstration

agriculture waste has been the most important method for heating and cooking. Dry biomass can also be burned to produce electricity, gasified to produce methane, hydrogen, and carbon monoxide, or converted to a liquid fuel. The wet form of biomass, such as sewage sludge, cattle manure, and food industry waste, can be fermented to produce fuel and fertilizer. Because biomass can be converted directly into a liquid fuel, it could supply much of our transportation fuel needs in the future for cars, trucks, buses, airplanes, and trains.

Anaerobic digestion is a biological process that converts solid or liquid biomass to a gas (a mixture of methane and CO_2) in the absence of O_2. Wet biomass from a variety of sources can be used, for example, those of industrial, agricultural, and domestic origins. The produced gas has heating value due to its methane content and is used for the generation of heat and electricity. An additional benefit from anaerobic digestion is that the solid and liquid residues can be used as compost and fertilizers (Bauen, Woods, and Hailes, 2004).

The most commonly used biofuel, ethanol, is currently produced from the food portion of crops (sugarcane, corn, and other grains). To reduce the pollution from automobile exhausts, ethanol is added to gasoline in many cities. The use of food items for biofuels may bring the risk of shifting food chains and also rising food prices. To address this problem, nonfood plant resources and the inedible parts of food crops (cellulosic material) are being seen as the future feedstock for biofuels. The recalcitrance (i.e., high resistance to decomposition or alteration) of cellulosic materials is mainly responsible for the poor efficiency of cellulose to ethanol conversion. Hence, ethanol produced from biomass at present is more expensive

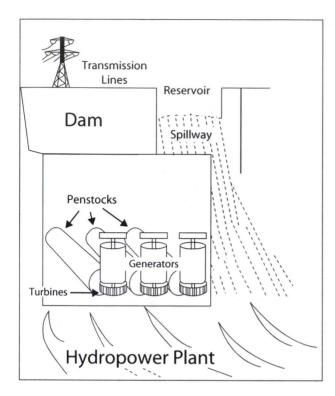

Figure 3.3. A typical hydropower plant (EIA, 2009b).

than gasoline. Research efforts are underway to produce cost-effective ethanol from grasses, trees, bark, sawdust, paper, and farming wastes. The next commonly used biofuel is biodiesel, which is produced from vegetable oils (e.g., soybean, corn, cottonseed, peanut, sunflower, palm, jatropha, or canola oils). However, the cost of biodiesel has increased because of the increased cost of vegetable oil, which is due to the competition with other edible oils (except jatropha oil, which is not edible). Efforts are underway to produce diesel from the synthesis gas produced by biomass gasification.

3.3 Hydropower

Energy from water velocity or pressure (head) in rivers and streams can be captured and turned into hydropower, also called hydroelectric power (Figure 3.3). Large-scale hydropower provides about 25% of the global electricity supply. In fact, more than 40% of the electricity used in the developing world is now supplied by hydropower. Energy supply from hydropower can be significantly increased; currently, of the technically usable 2200 gigawatts (GW) of large-scale hydropower, only 25% is being utilized.

Three key elements of hydropower are dams, turbines, and generators. Globally, dams cover nearly 500,000 km^2 of land and can store 6,000 km^3 of water. Due to the construction of dams, a small but measurable change in the distribution of fresh water has occurred in the world. The total installed capacity of large-scale hydropower is 612 GW and small-scale hydropower is 27.9 GW (Penche, 1998; Gleick, 1999). Hydropower plants are generally efficient in converting energy to electricity without any environmental pollution. However, objections are raised relative to the flooding of valuable real estate and scenic areas. Economic competitiveness of a particular hydroelectric installation depends on a number of factors, including fuel and construction costs. In general, hydropower resources are considered principal assets due to their high production potential and their economic efficiency.

Typically, energy efficiency for small turbines (10 kW) is about 60–80%, whereas that for the larger turbines is greater than 90%. The turbine shaft is attached to an electricity generator. The two main types of generator are synchronous and asynchronous generators. Both types of generators have been steadily improved over the years to provide high efficiency (98–99%). However, the older hydropower plants can have efficiency as low as 50%.

3.4 Geothermal

Geothermal energy is heat from the earth's interior, which has a temperature of 5,000–7,000°C, nearly as hot as the surface of the sun but substantially cooler than the sun's interior (13.6 million degrees Celsius). Low thermal conductivity, high thickness of the crust, and radioactive decay have kept the earth's core cooling to a very slow rate (about 1×10^{-7} °C/year) (Kutz, 2007). The earth's crust has anomalies (low thickness) at several locations that cause the molten core to venture close to the surface, which in turn causes formation of volcanoes, geysers, fumaroles, and hot springs (Figure 3.4).

Geothermal energy is inexpensive, nonpolluting, and renewable. Geothermal energy has been directly utilized in bathing, swimming and balneology (therapeutic mineral baths), space heating, greenhouses, fish farming, and industry (Fridleifsson, 2001). For example, in Southampton (UK) there is a district heating scheme based on geothermal energy, in which hot water is pumped up from about 1,800 meters below ground. The earliest recorded use of geothermal energy to produce electricity was in Lardarello (Italy) in 1904, with the commercial installation operational in 1913 for 0.25 megawatt electric (MWe). Since then, global use of geothermal energy has increased rapidly; it was recorded as 7,304 MWe in 1996. Geothermal resources have been identified in over 80 countries with record of use in 58 countries (Fridleifsson, 2001). Major countries with installed capacity are listed in Table 3.5.

Geothermal energy produces electricity with minimal environmental impact. With proven technology and abundant resources, geothermal energy can make a significant contribution toward reducing the emission of greenhouse gases. Table 3.6 compares the production costs of geothermal electricity with other energy sources from new plants in the United States, Japan, and New Zealand. Typically, the current electricity cost in US$/kWh is 0.02–0.10 for geothermal and hydro, 0.05–0.13

Table 3.5. *World's top countries using geothermal energy in direct uses (Fridleifsson, 2001)*

Country	Installed MWt (year 2000)
United States	3,766
China	2,282
Iceland	1,469
Japan	1,167
Turkey	820
Switzerland	547
Hungary	473
Sweden	377
France	326
Italy	326
New Zealand	308
Russia	308
Georgia	250
Mexico	164
Romania	152

Figure 3.4. Geothermal energy (INL, DOE, 2009)

Table 3.6. *Production costs of electricity from new plants, in US$/kWh (Murphy and Niitsuma, 1999)*

Country	Geothermal	Cheapest fossil fuel	Wind
United States	0.06	0.03 (natural gas)	0.06
Japan	0.06	0.07 (liquefied natural gas)	0.22
New Zealand	0.13	0.03 (various)	0.42

for wind, 0.05–0.15 for biomass, 0.25–1.25 for solar photovoltaic, and 0.12–0.18 for solar thermal electricity (Demirbas, 2006b).

3.5 Wind

Wind power has been used for centuries for water pumping and grain milling. More recently, since the 1980s, due to significant advances in aerodynamics, materials, design, and control of wind turbines, large wind turbines have been designed that are used to generate electricity (Kutz, 2007). Wind energy is nonpolluting and freely available in many areas of the world. In addition, wind turbines are becoming more efficient, thereby reducing the cost of electricity generated. Due to reduced equipment costs and government incentives, the wind energy sector is one of the fastest-growing energy sectors in the world with a growth rate of about 26%. For example, from 1991 to 2002, global installed capacity has increased from 2 GW to over 31 GW (Figure 3.5).

A major challenge is the erratic production of wind energy due to varying wind speed. Hence, it is more difficult to predict hourly wind power production and load, 1–2 days ahead of time, than to predict other production forms (Demirbas, 2006b). However, an advantage over solar power is that wind turbines can produce electricity 24 hours per day, whereas solar cells can only generate power during the daytime. Even in the windiest places, the wind does not blow all the time, so

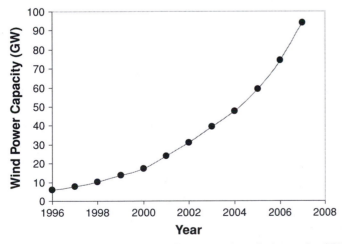

Figure 3.5. Cumulative global wind power installed capacity (GWEC, 2009).

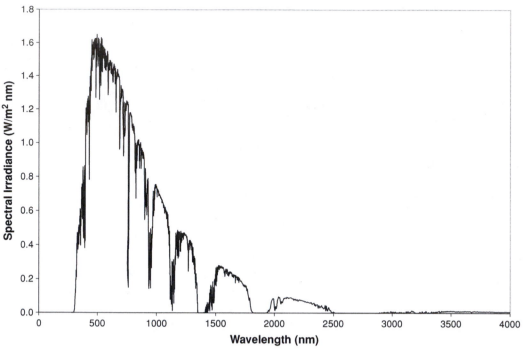

Figure 3.6. The solar spectral irradiance at 37° south facing (ASTM, 2003).

back-up batteries are needed for small wind systems. On the other hand, a large wind farm, without distributed wind turbines, can operate without needing batteries for back-up storage. In addition to the geographical and meteorological concerns, wind turbines do have some environmental concerns, including injury to birds, noise pollution, and aesthetics (Garg and Datta, 1998). Nonetheless the global wind power is expected to increase to about 210 GW by the year 2030 (Demirbas, 2006b).

3.6 Solar

The most abundant renewable energy resource on the earth is sunlight. A standard spectral irradiance on earth is shown in Figure 3.6; here, the area under the curve provides solar intensity in W/m^2 of the earth's surface. The magnitude of the solar flux for terrestrial applications varies with season, time of day, location, and orientation of the collecting surface (Kutz, 2007). For example, Figure 3.7 shows the variation of the solar flux in various locations in the United States.

Solar energy can be used in a variety of ways, including heating, cooking, drying, water heating, and electricity generation (via photovoltaic and thermal routes). Solar photovoltaic systems are being used for lighting, communication, and water pumping for drinking and irrigation (Garg and Datta, 1998). Other than photocells in photovoltaic systems, the major component of any solar system is the solar collector. Various types of collectors include flat-plate, compound parabolic, evacuated

Table 3.7. *Properties of common solar collectors (Demirbas, 2007)*

Type	Absorber	Motion	Temperature range (°C)
Flat plate collector	Flat	Stationary	27–77
Compound parabolic collector	Tubular	Stationary	57–237
Cylindrical trough collector	Tubular	Single-axis tracking	57–302
Parabolic trough collector	Tubular	Single-axis tracking	57–302
Parabolic dish collector	Point	Two-axes tracking	0–502
Heliostat trough collector	Point	Two-axes tracking	152–1987

tube, parabolic trough, Fresnel lens, parabolic dish, and heliostat field collectors; these are shown and compared in Table 3.7 (Kalogirou, 2004).

The current contribution of solar energy to the world energy supply is less than 1%. Similar to wind turbines, the cost of photovoltaic systems has fallen dramatically (Figure 3.8), and installation is on the rise at an annual rate of 30%. It is predicted that photovoltaic energy will be the largest renewable energy source by the year 2040 with production of 25% of the global power generation (EWEA, 2005). Although solar electricity is much more expensive than coal-based electricity (Table 3.8), it is expected that solar thermal power stations, based on parabolic

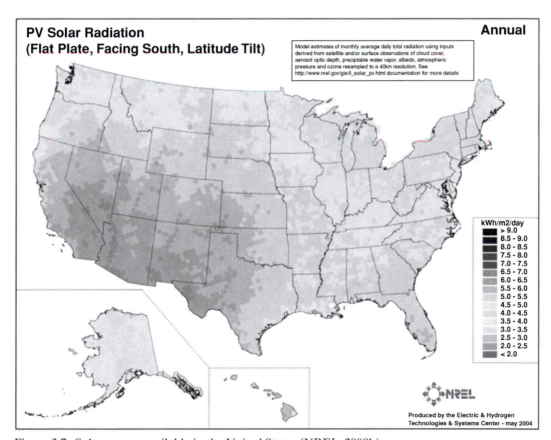

Figure 3.7. Solar energy available in the United States (NREL, 2009b).

Table 3.8. *Economics and emissions of conventional technologies compared with solar power generation (Demirbas, 2006b)*

Electricity generation technology	Carbon emissions (gC/kWh)	Generation costs (US cents/kWh)
Solar thermal and solar PV systems	0	9–40
Pulverized coal–natural gas turbine	100–230	5–7

and heliostat trough concentrating collectors, will soon become competitive in the world's electricity market (Trieb, 2000). This commercialization will benefit from the introduction of solar-supported power plants (SSPPs), which use technology similar to current coal-fired power plants. For example, SSPPs use similar steam turbine generators and fuel delivery systems. Concentrated solar energy is used to produce steam. The steam turns a turbine that drives a generator, producing electricity. In fact, in the transition, solar radiation can be used as either a primary or a secondary energy source to power gas turbines. Estimates show that the solar energy system has a net fuel-to-electricity efficiency higher than 60%, even when the energy to produce high-pressure oxygen and to liquefy the captured CO_2 is taken into account (Kosugi and Pyong, 2003).

3.7 Ocean Energy

Ocean energy includes wave energy, tidal energy, and ocean thermal energy. Theoretically, the world's wave energy resource is between 200 and 5,000 GW and is mostly found in offshore locations (Garg and Datta, 1998). Due to the proximity to the energy market, wave energy converters at the shoreline are likely the first to be fully developed and deployed, but waves are typically two to three times more

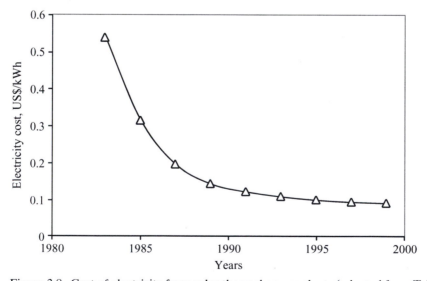

Figure 3.8. Cost of electricity from solar thermal power plants (adapted from Trieb, 2000).

Table 3.9. *Typical composition of raw landfill gas (Demirbas, 2008a)*

Component	Chemical formula	Content (volume %)
Methane	CH_4	40–60
Carbon dioxide	CO_2	20–40
Nitrogen	N_2	2–20
Oxygen	O_2	<1
Heavier hydrocarbons	C_nH_{2n+2}	<1
Hydrogen sulfide	H_2S	0.0040–0.0100
Complex organics	–	0.1–0.2

powerful in deep offshore waters than at the shoreline. In comparison to wind and solar energies, wave energy and tidal stream are very much in their infancy. In ocean thermal energy conversion (OTEC), the oceans' natural temperature gradient is used to drive a power-producing cycle. Typically, a difference of 20°C between the warm surface water and the cold deep water is enough to operate OTEC. Due to the vastness of the oceans, the potential for OTEC energy is very high. A typical design is shown in Figure 3.9.

3.8 Biogas

Biogas can be obtained from digesting the organic material of municipal solid wastes and animal manures. Biogas is composed of methane (CH_4), CO_2, air, ammonia, carbon monoxide, hydrogen, sulfur gases, nitrogen, and O_2. A typical analysis of raw landfill gas is shown in Table 3.9. Because of the widely varying nature of the

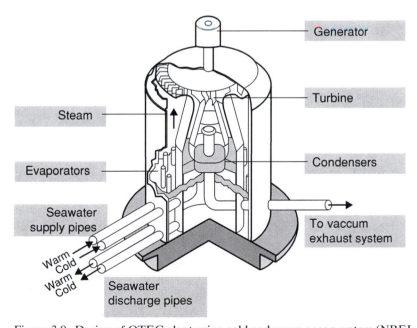

Figure 3.9. Design of OTEC plant using cold and warm ocean waters (NREL, 2009a).

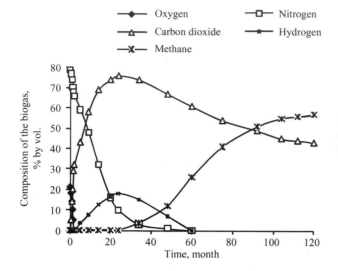

Figure 3.10. Production of biogas components with time in landfill (adapted from Demirbas, 2008a)

contents of landfill sites, the constituents of landfill gases vary widely (Demirbas, 2008a). Due to its energy value, CH_4 is the most important component of landfill gases. CH_4 makes up around 40–60% of the gas produced, which is a major contributor to global warming. Hence, it is a good practice to collect landfill gas and use it to create energy. During energy use, CH_4 is combusted to CO_2, which has a lower greenhouse gas effect than CH_4, and the CO_2 form of carbon can be easily used in the growth of new plants. In addition, collection and use of landfill gas improves local air quality and eliminates a potential explosion hazard.

The composition of biogas changes as the landfill site matures (Figure 3.10). The early stage is characterized by the removal of O_2 from the waste by aerobic bacteria.

In the mid-stage, a diverse population of hydrolytic and fermentative bacteria hydrolyzes cellulose, hemicelluloses, proteins, and lipids into soluble sugars, amino acids, long-chain carboxylic acids, and glycerol. In the later stage, biogas is produced from soluble molecules.

3.9 Summary

Biomass, hydropower, geothermal, wind, and solar energies have been used for cooking, heating, milling, and other tasks for thousands of years. However, these forms of energies were replaced by coal and petroleum during the Industrial Revolution in the eighteenth century. The high energy density of coal and petroleum was of significant help in rapid industrialization. Because of concerns about fossil fuel depletion and environmental pollution, attention has focused on renewable energy since the 1970s. Efforts have led to experiments with solar steam for industry and solid wood, methanol gas, or liquid biofuels for engines. In recent decades, efforts for a shift toward renewable energy have intensified due to the concern of CO_2 emission from fossil fuels contributing to global warming. In contrast with fossil fuels, renewable fuels are available around the world, cleaner, environmentally benign, and far more abundant. However, they are also more dispersed, more expensive to collect, and intermittent in nature, requiring a more expensive production system and storage. However, some forms of renewable fuel are expected to be cost competitive with fossil fuel in the near future.

4 Biomass Availability in the World

4.1 Introduction

Plants are natural solar cells that capture sunlight and store it as carbohydrate, consuming water and carbon dioxide (CO_2) and releasing oxygen (O_2) in the process (Figure 4.1).

$$CO_2 + 2H_2O + \text{light and heat} \rightarrow [CH_2O] + H_2O + O_2$$

The energy use of plants (e.g., combustion) is the reverse of photosynthesis. Terrestrial plants only capture 0.02% of solar energy reaching the earth. However, due to the large surface area of the earth, this amount of energy is equivalent to about seven times the energy consumption by the world population. Typically, one square meter of land can store $\sim$18 megajoules (MJ) per year of the growing crop. Some plants do not produce any food, whereas other plants produce both food and nonfood components. The nonfood component contains lignin, cellulose, and hemi-celluloses, which is referred to as "lignocellulosic biomass."

Biomass is converted to energy by direct combustion or by conversion to gas or liquid fuel, which is then followed by combustion. The product of biofuel use is CO_2, which fits with the natural carbon cycle (Figure 4.2).

4.2 Biomass Definition

Biomass usually refers to wood, grass, short rotation woody and herbaceous crops, bagasse, wood waste, sawdust, agricultural waste, industrial residues, waste paper, municipal solid wastes, waste from food processing, aquatic plants, algae, animal waste, and similar materials with energy value. In theory, fossil fuels can also be termed biomass because they are the fossilized remains of plants that grew millions of years ago. However, in practice, fossil fuels are excluded from the biomass definition, mainly because the use of fossil fuels causes an increase in CO_2 concentration in the atmosphere. On the other hand, the use of biomass as fuel does not increase CO_2 concentration in the atmosphere; the CO_2 released during combustion of biofuel is removed from the environment by photosynthesis during the production of

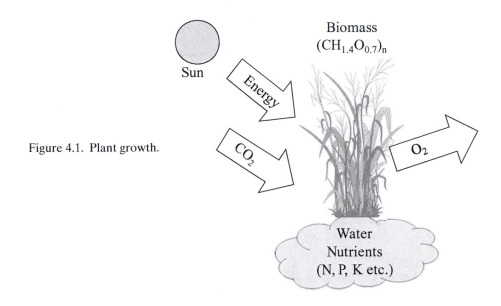

Figure 4.1. Plant growth.

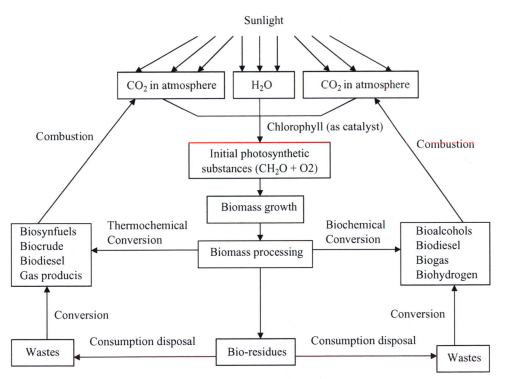

Figure 4.2. Carbon cycle, photosynthesis, and main steps of biomass technologies.

Table 4.1. *Biomass energy use in different regions of the world (Demirbas, 2008a)*

Region	Share of biomass in final energy consumption (%)
Africa	60.0
South Asia	56.3
East Asia	25.1
China	23.5
Latin America	18.2
Europe	3.5
North America	2.7
Middle East	0.3

biomass. Chemical constituents of biomass are cellulose, hemicelluloses, lignin, and ash, with minor amounts of extractives, lipids, proteins, and starch.

4.3 Biomass Sources

Commercial biomass sources are already developed in the world; biomass ranks fourth as an energy resource, providing approximately 14% of the world's energy needs. However, biomass use has a high regional variation (Table 4.1). The use of biomass is highest in Africa, Asia, and Latin America. For large portions of the rural populations of developing countries, and for the poorest sections of urban populations, biomass is often the only available and affordable source of energy for basic needs, such as cooking and heating. In some countries (e.g., Angola, Ethiopia, Mozambique, Tanzania, Democratic Republic of Congo, Nepal, and Myanmar), biomass provides as much as 80–90% of the energy.

4.3.1 Solid Wastes

Wastes from industrial and municipal sources present an attractive biomass source, as the material has already been collected and can be acquired at a negative cost due to tipping fees (i.e., sources will pay money to get rid of waste). Municipal solid waste (MSW) is defined as waste durable goods, nondurable goods, containers and packaging, food scraps, yard trimmings, and miscellaneous inorganic wastes from residential, commercial, and industrial sources (Demirbas, 2004b). A typical composition of MSW is shown in Table 4.2, and the nonmetallic/glass portion is responsible for a heating value of 12.4 MJ/kg moisture-free basis. Availability of MSW near population centers is another positive feature. It is estimated that, in the United States alone, 24–30 million dry tons of MSW are available each year for biofuel production.

Until the 1990s, combustion technology was the main driver when dealing with MSW, with the objective of waste treatment for disposal and, in some cases, sorting. In the 1990s, environmental and socioeconomical concerns supported the direct

Table 4.2. *Composition of domestic solid waste (Demirbas, 2004b)*

Component	Composition (wt%)
Paper waste	33.2–50.7
Food waste	18.3–21.2
Plastic matter	7.8–11.2
Metal	7.3–10.5
Glass	8.6–10.2
Textile	2.0–2.8
Wood	1.8–2.9
Leather and rubber	0.6–1.0
Miscellaneous	1.2–1.8

recycling of wastes and residue. However, the recycling approach has been limited due to concerns about costs and the presence of pollutants (e.g., heavy metals, organics) in MSW. However, now with the higher energy cost, transformation of MSW to valuable materials and energy (i.e., valorization) is emerging as a strong trend.

4.3.2 Agriculture Residue

Major agricultural residues includes crop residue, straws and husks, olive pits and nut shells, and so on (Figure 4.3). The residue can be divided into two general categories: (1) field residue: material left in the fields or orchards after the crop has been harvested, and (2) process residue: materials left after the processing of the crop into a usable resource. The field residues include stalks, stems, leaves, and seed pods. The process residues include husks, seeds, bagasse, and roots. Some agricultural residues are used as animal fodder, for soil management, and in manufacturing.

4.3.2.1 Cereal Straw
Cereal straw is the dry stalk of a cereal plant that is left behind in the field after the grain or seed has been removed during combining. A typical cereal crop (e.g.,

Figure 4.3. Crop residues, straws, and husks (adapted from CleanTechnica, 2008).

wheat) produces about 2.5–5 tons per hectare (ha) of straw, with a straw-to-grain ratio of ~1.5. After harvesting, a small amount of straw is used as feed, animal bedding, mushroom compost, or garden mulch, and the majority is burned or incorporated in the soil. Field burning has been practiced as a means of straw disposal, but now many countries are putting bans on this practice due to the pollution generated. The collection cost (raking, baling, etc.) of straw is around $25/ton. The volumetric energy density of straw is about 10 to 20 times less than that of coal or petroleum. Hence, the compaction technologies have been developed to produce straw pellets and wafers with a high mass density, but the cost of such compaction is high.

4.3.2.2 Corn Stover

Corn stover is the nongrain and above-ground part of the corn plant consisting of stalk, tassel, leaves, cob, husk, and silk. On average, 1 ha (10,000 m^2) can produce seven tons of grain and six tons of stover. The crown and its surface roots are not considered part of the stover. Currently, about 5% stover is used for animal bedding and feed, and the remaining is plowed into the soil. Corn stover has potential for direct burning, conversion into biofuels, and fiber for pulp and paper industry as well as the particle board and oriented strand board industries (Savoie and Descoteaux, 2004). It is estimated that 94.6 million dry tons per year of corn stover are available for biofuel production in the United States.

4.3.2.3 Rice Husk

Due to its widespread production, rice husk is the world's most common residue. One ton of husk is produced for four tons of rice grain produced. Globally, 100 million tons of rice husk are produced each year, and it is mostly used as bedding material for animals. The majority of production is concentrated in Western and Eastern Asia (India, China, Japan, Indonesia, Thailand, Burma, and Bangladesh). Rice husk is removed from the grain in the processing plants; hence, this biomass is available in the collected form. In addition, rice husk is uniform in nature and has a better flow characteristic than other forms of biomass. Hence, biofuel technologies such as gasification can use husk, as uniform fuel quality is needed for the best results. However, husk has a relatively high silica content, which can cause ash and slag problems in the boiler on the combustion.

4.3.2.4 Bagasse

Sugarcane has one of the best photosynthesis efficiencies as it can convert up to 2% of incident solar energy into biomass. The cultivation requires a tropical or subtropical climate with a minimum of 600 mm of annual rainfall. In addition, the cultivation does not require too much input of fertilizer or pesticide. Each ton of sugar cane (burned and cropped) produces 135 kg of sugar and 130 kg of dry bagasse. The global production of bagasse is 200 million dry tons each year, with major production in Brazil, India, China, and Thailand. The annual production varies from year to year, with a cane crushing period of 6–7 months. Bagasse has a heating value of 19.2 MJ/kg-dry basis. Most sugar mills burn bagasse to produce heat needed for cooking of cane and for the evaporation of water from the syrup. But, usually, there is a large excess of bagasse at the plant, and, traditionally, mills have resorted to inefficient burning as a method of disposal. Recently, some mills have started generating electricity that is sold for external use. With better value from biofuels, bagasse is expected to become an attractive feedstock.

4.3.3 Energy Crops

There are a number of plants that can convert solar energy into biomass at a high efficiency, including herbaceous woody crops, short rotation woody crops, forage crops (alfalfa, clover, switchgrass, and miscanthus grasses), sugar crops (sugar cane, sugar beet, fiber sorghum, and sweet sorghum), starch crops (corn, barley, and wheat), and oil crops (soybean, canola, palm, sunflower, safflower, rapeseed, and cotton). The annual production from each hectare of land is equivalent to about 400 GJ for C4 crops, 250 GJ for starch crops, and 70 GJ for oil crops. Hence, high-yield energy crops are of interest for biofuel production. However, in some locations, the residue from food crops may be cheaper due to the high revenue from the sale of grain, oil, or sugar. Nonetheless, the focus is to raise an energy crop that requires less input (fertilizer, water, pesticide), does not compete with the food supply, and can use low-value marginal land. Efforts are underway to genetically modify the crop to increase the yield and reduce input cost.

Woody crops (e.g., willow, poplar) and tropical grasses (e.g, napier grass, elephant grass) are receiving more attention from energy crop companies. Woody plants grow slowly and have hard external surfaces, whereas herbaceous plants grow faster with loosely bound fibers due to lower lignin content. In fact, the relative proportion of cellulose and lignin is one of the determining factors in identifying the suitability of plant species for subsequent processing as energy crops. Both woody and herbaceous plants require specific growing conditions, including soil type, soil moisture, nutrient balances, and sunlight; all of these factors determine their suitability and production rate. Table 4.3 summarizes the biomass yields from selected perennial crops from recent field studies (Sanderson and Adler, 2008).

Switchgrass (*Panicum virgatum*) is a perennial warm season grass native to North America (Figure 4.4). It naturally occurs southward of 55°N in Canada, the United States, and Mexico. Traditionally, it has been used for forage production, soil conservation, game cover, and as an ornamental grass. The use for heat and fiber is more recent, with interest in ethanol production being the most recent. The ethanol production of as high as 100 gallons (380 liters) per ton is being proposed. With a yield of 8 tons/ha of switchgrass, one can expect to produce 800 gallons of ethanol/ha. Coupled with high ethanol production, hardiness in poor soil and climate conditions, rapid growth, and low fertilization and herbicide requirements, switchgrass is an attractive route to produce biofuels.

Sorghum consists of numerous subspecies (Figure 4.5). Many sorghum species are used for food, such as grains and syrups. Some species are used as fodder plants, either cultivated or part of a pasture. In fact, sorghum species are important cereal

Table 4.3. *Example biomass yields from selected perennial crops (Sanderson and Adler, 2008)*

| Crop | Yield | | | |
	Biomass range (tons/ha)	Mean energy (GJ/ha)	Nitrogen fertilizer input (kg N/ha)	Location and description
Switch grass	5.2–11.1	134	0–212	Field-scale plots (3–9.5 ha) on 10 farms in Nebraska, South Dakota, and North Dakota, harvested for 5 years
Miscanthus	1.4–18.2	150	60	Experimental plots harvested for 3 years in Denmark; extensive work done at the University of Illinois at Urbana–Champaign
Miscanthus	7.5–40.9	416	60	Experimental plots irrigated and harvested for 3 years in Portugal
Reed canary grass	9.4–10.1	165	0–168	Experimental plots in Indiana harvested for 3 years
Reed canary grass	5.5–10.2	127	140	Experimental plots at two sites in Iowa harvested for 5 years
Alfalfa	7.0–12.0	157	0	Experimental plots at two sites in Minnesota harvested for 2 years
Bermuda grass	12.8–19.9	248	Not reported	Three experimental plot sites in Georgia harvested for 3 years
Eastern gama grass	6.5–15.9	185	84–301	Summary of studies from nine states in the Eastern United States
Prairie cord grass	4.6–8.6	106	0	Experimental plots in South Dakota harvested for 4 years

crops in Central America, Africa, and South Asia. The juice from sorghum stalks can be directly fermented to produce ethanol fuel. In the United States, tests are underway to identify the best varieties to provide a high ethanol yield from sorghum leaves and stalks.

The oil palm is a tropical tree originally found growing wild in West Africa. Later it was developed into an agricultural crop in Malaysia (Figure 4.6). Oil palm plants can produce about 4–5 tons of oil/ha/year, which is about 10 times the yield of oil from soybean. Oil palm is the highest oil-producing plant on a per acre basis. Due to the phenomenal growth of the palm cultivation, Malaysia is now the largest producer and exporter of palm oil in the world. This palm oil export accounts

Figure 4.4. Switchgrass in the United States (U.S. DOE, EFRE, 2009).

Figure 4.5. Sorghum (NBII, U.S. Geological Survey, 2009).

(a)   (b)   (c)

Figure 4.6. (a) Palm tree, (b) palm fruit, and (c) palm biomass (Sumathi, Chai, and Mohamed, 2008; reproduced with permission of Elsevier).

for 52% (26.3 million tons) of total world oils and fat export. As a byproduct of oil production, a good amount of biomass is now available from the palm oil industry, which can be used for energy applications.

Jatropha, a nonedible oil plant, is gaining prominence. The Jatropha curcas variety is suitable for oil production, as its seeds contain up to 37 wt% oil. It is a drought-resistant perennial and grows well in marginal/poor soil. The plant can keep producing seeds for up to 50 years. Seeds need to be harvested at maturity, and not all fruits mature at the same time. The seed harvesting is done over an extended period (about 2 months in semi-arid regions and year-long in permanently humid regions) and is labor-intensive; hence, Jatropha plantation provides an excellent opportunity for rural employment in the developing countries. The actual mature seed yield is not accurately established. Wide-ranging annual yields have been reported from 0.4 to 12 tons/ha (Achten et al., 2008; Openshaw, 2000).

4.3.4 Pulp and Paper Industry Waste

Operations in the pulp and paper industry generate a vast amount of biomass residue, including bark, leaves, needles, branches, and sludge. The nonsludge biomass can easily be used in one of the biofuel production processes. Pulp mill sludge represents an attractive feedstock for biofuels, as currently most of the sludge is disposed in landfills where it degrades to methane gas, a more harmful greenhouse gas than carbon dioxide. In the United States alone, 4–5 million tons/year pulp mill sludge is available. By using it in biofuel production, land and water contamination from landfills can be avoided.

Another byproduct is black liquor, which contains a significant energy value due to the dissolved lignin. Currently, black liquor is combusted to produce needed energy in the paper mill, and the extra energy is used to raise electricity for export. Black liquor composition can be represented as $C_{10}H_{12.5}O_7Na_{2.4}S_{0.36}$ (Backman, Frederick, and Hupa, 1993; Salmenoja, 1993), and its components with energy value

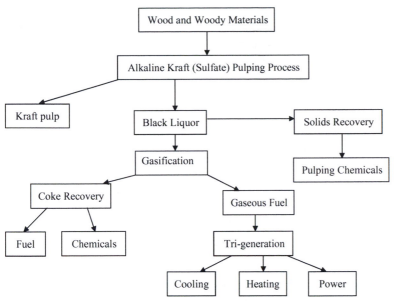

Figure 4.7. Flow diagram of alkaline kraft pulping process and black liquor gasification for energy recovery.

mainly consist of dissolved lignin degradation products along with the cellulosic and hemicellulosic hexose and pentose sugar degradation products (Demirbas, 2002b). Black liquor has a solid content of approximately 15% by weight. To raise the solid content in the liquor, it is concentrated to about 55–75% solids and burned in old Tomlinson-type recovery boilers and alternative combustion systems that are being developed (Kohl, 1986). Apart from the chemical recovery, black liquors can be used to produce liquid, gas, and char fuels. Black liquor gasification offers pulp and paper mills an efficient means to recover biomass energy. Figure 4.7 shows the flow diagram for the alkaline kraft pulping process and black liquor gasification for energy recovery. Upon gasification, organic fraction of the black liquor is converted into a clean gaseous fuel suitable for use in a gas turbine (Salmenoja, 1993). Here, organic components are converted into hydrogen (H_2), methane (CH_4), and carbon monoxide (CO) gases.

4.3.5 Wood and Forest Waste

Forest harvesting, a major source of biomass for energy, is carried out to thin young stands, to cut old stands for timber/pulping, or to remove the stands damaged by insects, disease, or fire. The tops and branches are obtained that have little timber or pulp value but have good energy content. Typical forest harvesting removes only 25–50% of volume, leaving the remaining volume as biomass for energy use. Forest residues normally have low density and fuel values that keep transport costs high, and so it is economical to reduce the biomass density in the forest itself.

Forest management to avoid forest fires also represents an attractive source of biomass. Each year in the United States alone, 30,000 fire incidents are reported,

Table 4.4. *Typical forest tree biomass (Nakamura and UC Cooperative Extension, 2004)*

	Biomass Tons/acre	Mass (%)
Foliage	16.3	8
Branches	16.3	8
Bark	16.3	8
Stem wood	73.1	33
Tree total (above ground)	121.7	57
Forest floor	32.2	15
Soild (up to 3 feet deep)	60.8	28
Forest total	214.7	100

covering about 1 million acres of forest (NIFC, 2009). Biomass available in typical forest land is shown in Table 4.4. One million acres of forest fires per year represent 214.7 million tons of biomass or 3.4×10^9 gigajoules (GJ) of energy (enough to fuel 34 million cars per year, with each car needing 100 GJ/year). In silviculture (the science and art of producing and tending a forest), thinning of forest trees is a basic tool in which trees that are growing too close together are removed. Thinning can be an effective tool in reducing wildfires (Figure 4.8).

4.3.6 Algae

Due to the high sunlight-to-biomass conversion efficiency, algae are gaining attention for fuel use. Microalgae are the fastest-growing photosynthesizing organisms, as they can complete an entire growing cycle within only a few days. In some algae, up to 50% of their mass contains oil content. Hence, production of algae oil for biodiesel is gaining attention in experimental efforts. However, due to the high cost involved in the separation of algae from water, there has not been any commercial undertaking so far. It is estimated that 40 tons of oil/ha/year can be produced using diatom algae, which is 7 to 31 times more than the yield from the best-performing vegetable oil plant, oil palm, and 200 times higher than the soybean plant (Sheehan et al., 1998).

Figure 4.8. Forest before and after thinning (NRCS, USDA, 2009).

Table 4.5. *Biomass potential and use distribution between regions, in billion tons (Smeets and Faaij, 2007)*

Biomass potential	United States + Canada	Latin America	Asia	Africa	Europe	Middle East	Former USSR	World
Woody biomass	0.80	0.37	0.48	0.34	0.25	0.03	0.34	2.60
Energy crops	0.26	0.76	0.07	0.87	0.16	0.00	0.23	2.34
Straw	0.14	0.11	0.62	0.06	0.10	0.01	0.04	1.08
Other	0.05	0.11	0.18	0.08	0.04	0.01	0.02	0.48
Total sustainable potential	1.24	1.34	1.35	1.34	0.56	0.04	0.63	6.49
Use	0.19	0.16	1.45	0.52	0.13	0.00	0.03	2.48
Excess biomass (total potential – use)	1.05	1.18	−0.1	0.82	0.43	0.04	0.6	4.01

Notes: A heating value of 16 MJ/kg is used. Asia uses 0.1 billion tons/year more than the currently sustainable biomass; however, with improved efficiency, biomass availability may increase.

4.4 World Potential to Product Biomass

Biomass can be considered the best option of renewable energy that can seamlessly be coupled with the existing liquid petroleum fuel network. Biomass potential for energy use in various parts of the world is shown in Table 4.5.

The current annual world biomass potential is 6.49 billion tons, out of which 2.48 billion tons are already being used. The excess biomass, 4.01 billion tons, can be used for modern biofuel production. The world's total forest area is estimated to be about 4 billion hectares (30% of the total land area), which corresponds to a global average of 0.6 ha of forest per capita. Most of the world's forest land is located in the Russian Federation, Brazil, Canada, the United States, and China, with the Russian Federation accounting for 20% of the forest land. The excess capacity is mostly available in the Americas and Africa. Other than in Asia, biomass in the rest of the world remains underutilized. The efficiency of biomass use in Asia could be improved to increase that continent's biomass availability. For example, traditional biomass-fired stoves have only 6–12% energy efficiency. With the use of inexpensive modern stoves, this efficiency could be improved two- to threefold. Amounts of biomass used for cooking in different parts of Asia are shown in Table 4.6, for a total of nearly 392 million tons per year. With an increase in stove efficiency, the same cooking could be performed using only 157 million tons, which would leave 235 million tons for modern biofuel conversion.

4.5 Biomass Characterization

The key character of biomass can be quantified in terms of particle size, bulk density, moisture content, ash content, heating value, molecular composition (e.g., lignin, hemicelluloses, cellulose, volatiles, and protein), and elemental composition (e.g., C, O, H, N, S, K, and P). The American Society for Testing and Material's

Table 4.6. *Biomass consumption using traditional stoves in Asia, in million tons/year (Bhattacharya and Abdul Salam, 2002)*

Country	Fuelwood	Agriculture residue	Animal dung	Total
China	24.8	52.5	10	87.3
India	99.9	31.7	63.8	195.4
Nepal	10.1	3.0	1.8	14.9
Pakistan	27.0	8.05	13.3	48.35
Philippines	11.64	2.64	0	14.28
Sri Lanka	4.48	0.99	0	5.47
Vietnam	20.83	5.23	0	26.06
Total	198.75	104.11	88.9	391.76

ASTM 870–82 standard test method has been developed for analysis of wood fuel. The U.S. Department of Energy's online database discusses properties of various types of biomass. This database can be accessed at the following Web site: http://www1.eere.energy.gov/biomass/feedstock_databases.html.

The standard protocols for various analyses are listed in Table 4.7.

Table 4.7. *Standard protocols for biomass characterization*

Analysis	Standard protocol
Acetic acid	LAP-017
Acid-soluble lignin	LAP-004
Acid-insoluble lignin	ASTM E-1721–95
Arabinan	ASTM E-1821–96 or E-1758–95
Ash	ASTM D-1102
Cellulose	ASTM E-1821 or E-1758–95
Extractives	ASTM E-1690–95
Galactan	ASTM E-1821–96 or E-1758–95
Glucan	ASTM E-1821–96 or E-1758–95
Hemicelluloses	ASTM E-1821–96 or E-1758–95
Mannan	ASTM E-1821–96 or E-1758–95
Starch	LAP-016
Total lignin	ASTM E-1721–95 and T-250 (or LAP-004)
Uronic acids	Scott 1979
Xylan	ASTM E-1821–96 or E-1758–95
C	ASTM E-777
H	ASTM E-777
N	ASTM E-778
O	By difference
S	ASTM E-775
Volatile matter	ASTM D E-872
High heating value, moisture-free	ASTM E-711
Lower heating value, moisture-free	ASTM D-2015

Table 4.8. *Characteristics of typical biomass (ORNL, biochar, 2009)*

Biomass	Ash (%)	Sulfur (%)	Potassium (%)	Heating value (MJ/kg)
Miscanthus	1.5–4.5	0.1	0.37–1.12	17.1–19.4
Switchgrass	4.5–5.8	0.12		18.3
Hybrid poplar	0.5–1.5	0.03	0.3	19.0
Softwood	0.3	0.01		19.6
Hardwood	0.45	0.009	0.04	20.5
Sugarcane bagasse	3.2–5.5	0.10–0.15	0.73–0.97	18.1
Sweet sorghum	5.5			15.4
Corn stover	5.6			17.6
Bituminous coal (for comparison)	1–10	0.5–1.5	0.06–0.15	27–30

4.5.1 Moisture Content

Moisture in biomass is stored in spaces within the cells and cell walls. Moisture content varies from one part of the tree to another. It is often the lowest in the stem and increases toward the roots and the crown. When biomass is removed, water respiration continues for some time, along with the removal of moisture. Hence, harvested biomass is left in the field for some time to automatically reduce the moisture content. When biomass is dried and stored, the moisture equilibrates with the ambient air humidity. Hence, the air-dried biomass typically contains 20% moisture. Moisture content (wt%) can be defined on wet or dry basis as follows.

Moisture content (wet basis) = (total weight of wet wood
− over-dry weight)/(total weight of wet wood) × 100

Moisture content (dry basis) = (total weight of wet wood
− over-dry weight)/(over-dry weight) × 100

4.5.2 Ash Content

Ash or inorganic content in biomass depends on the type of plant and the soil it grows in. For example, the ash contents of hard and soft woods are about 0.45 and 0.3 wt%, respectively (Table 4.8). Ash content can increase during harvesting due to soil contamination of the biomass. Ash sintering and melting temperatures are also important for design of the processes that use high temperatures (e.g., gasification and pyrolysis). For biomass, typical ash sintering temperature is around 600–900°C and melting temperature is around 1000–1300°C.

Biomass contains higher alkali metals compared with coal. Sodium and potassium compounds become easily volatile during combustion and redeposit on cooler surfaces (e.g., heat exchanger), which causes blockage. Also, cofiring of biomass and coal can cause slagging, even when combustion of the individual components may be satisfactory.

4.5.3 Heating Value

For biofuel application, the energy content of biomass is the most important factor. The heat value of any carbonaceous fuel depends on the oxidation state of the carbon atoms. The higher the oxidation state, the lower the heating value. For example, heating value is in the order $CH_4 > CH_4O > CO > CO_2$. In fact, the heating value of CO_2 is zero, as after the complete combustion CO_2 is the product:

$$CH_{1.4}O_{0.7} + O_2 \rightarrow CO_2 + 0.7H_2O + heat$$

The higher heating value (HHV) or gross heating value includes the latent heat of condensation as the water vapors are allowed to condense. Demirbas et al. (1997) have developed a formula for estimating HHV (in MJ/kg) based on Dulong's formula:

$$HHV = 0.335C + 1.423H - 0.154O - 0.145N$$

which uses wt% carbon (C), hydrogen (H), oxygen (O), and nitrogen (N) values. The heating values of biomass components cellulose, hemicelluloses, and lignin are 18.6, 18.6, and 25 MJ/kg, respectively. Hence, as the relative ratios of cellulose, hemicelluloses, and lignin change from one biomass to another, the heating value changes. In addition, the volatiles and extractives will also affect the heating value of biomass. Nonetheless, the above formula for HHV can be used for all types of biomass.

4.5.4 Organic Chemical Composition

The organic chemicals in biomass include volatiles, lignin, cellulose, hemicelluloses, and extractives. The relative composition of the components depends on the source of the biomass (Table 4.9). The composition is also represented in terms of elemental analysis in the form of C, N, O, and H content.

Cellulose is a glucose polymer, consisting of linear chains of (1,4)-D-glucopyranose units, in which the units are linked 1–4 in the β-configuration, with an average molecular weight of around 100,000. Hemicelluloses are a mixture of polysaccharides with a typical molecular weight of 30,000 or less and are composed of glucose, mannose, xylose, and arabinose sugars and methylglucuronic and galacturonic acids. Hemicelluloses, in contrast to cellulose, are heterogeneous branched polysaccharides that bind tightly, but noncovalently, to the surface of each cellulose microfibril. Lignin is an amorphous, high-molecular-weight, cross-linked polymer made of phenyl propane units. The monomer unit may have none, one, or two methoxyl groups attached to the aromatic ring, giving rise to three lignin structures of type I, II, and III, respectively. Grasses are usually rich in type I lignin, conifer woods are rich in type II lignin, and deciduous woods are rich in type III lignin.

Table 4.9. *Organic chemicals in biomass (Kong, Li, and Wang, 2008)*

	Maize straw	Saw dust	Rice husk	Wheat bran
Chemical composition (mass %, moisture-free)				
Ash	3.4	5.3	15.5	3.7
Lignin	17.3	30.3	20.8	21.4
Cellulose	24.1	12.5	23.7	38.6
Hemicelluloses	28.6	28.1	45.3	19.3
Extractives	30.0	29.1	10.7	17.1
Elemental analysis (mass %, ash-free)				
Carbon	44.1	43.4	32.4	42.2
Nitrogen	1.2	2.2	1.1	3.6
Oxygen	47.4	47.0	59.5	46.0
Hydrogen	7.3	7.4	6.9	7.9

4.5.5 Density

Biomass, especially originating from agriculture, exhibits a low bulk density (Table 4.10). Hence, transportation cost is a concern when the biomass needs to be taken far away from the biofuel plant. Also, forest residue (leaves and twigs) presents the same problem. However, the high surface area presented by straws, leaves, and twigs is advantageous in the conversion process, as the reaction front travels through the surface of the biomass particle.

Density can be increased by biomass densification or compaction processes in which small particles of biomass (e.g., straw, sawdust, chips, twigs) are pressed into small pellets or briquettes. The density can be increased as much as tenfold of the original density. However, for such a compaction to work, the moisture content of the biomass should be between 7% and 14%. For higher moisture content, the biomass does not compact easily, and for the lower moisture content, the particles do not bind effectively. Due to the popularity of the pellet boiler systems in Sweden, Austria, and Germany, pelletizing machines and industry are well developed. These boilers use pellets made primarily from the biomass waste leftover (e.g., shavings and sawdust) from processing of trees for lumber (Figure 4.9).

Table 4.10. *Bulk density and moisture content of typical biomass (Sims, 2002)*

Type of biomass	Bulk density (kg/m³)	Moisture content (% wet basis)
Green roundwood	510–720	40–50
Green wood chips	280–410	40–50
Air-dried roundwood	350–530	20–25
Air-dried wood chips	190–290	20–25
Kiln-dried wood chips	160–250	10–15
Straw bales	200–500	10–15
Coal for comparison	700–800	6–10

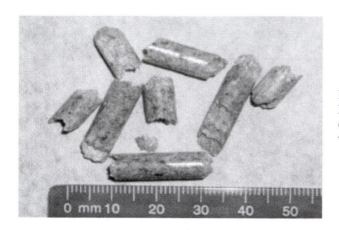

Figure 4.9. Picture of softwood pellets. Note: Ruler scale is in millimeters (Hartley and Wood, 2008; reproduced with permission from Elsevier).

4.6 Summary

Biomass availability in the world is more diverse than that of fossil fuel. Presently, 4 billion tons/year excess biomass is available worldwide for conversion to biofuels. For the world population of 6.7 billion, this represents biomass of 597 kg/person/year or 11 GJ/person/year. For comparison, the world use of petroleum is at 29 GJ/person/year. Currently, most of the biomass is available as waste residue from forest and agriculture industries. Additional biomass can be produced by raising energy crops on marginal lands and from forest management (i.e., to avoid forest fires). The properties of biomass can vary greatly from the types of species, location and time of harvesting, and the harvesting process. The key character of biomass can be quantified in terms of particle size, bulk density, moisture content, ash content, heating value, molecular composition (e.g., lignin, hemicelluloses, cellulose, volatiles, and protein), and elemental composition (e.g., C, O, H, N, S, K, and P). The high moisture content and low bulk density of some of the biomass are of concern, as they increase transportation and processing costs.

5 Conventional Ethanol Production from Corn and Sugarcane

5.1 Introduction

5.1.1 Ethanol

Ethanol – also known as ethyl alcohol, pure alcohol, grain alcohol, alcohol, spirit, and hydroxyethane – is a flammable and colorless liquid with a boiling point of 78.4°C, melting point of −114.3°C, and density of 0.79 g/cm³. Ethanol is used in alcoholic beverages, solvents, scents, flavorings, coloring, medicines, chemical synthesis, and thermometers. With its molecular formula of C_2H_5OH, ethanol contains 52 wt% carbon, 13 wt% hydrogen, and 35 wt% oxygen. Due to its heating value, ethanol has a long history of use as a fuel for heating and lighting; recently, it has been used as a fuel for internal combustion engines. The fermentation of sugar into ethanol is one of the earliest organic reactions employed by humanity. In modern times, ethanol intended for industrial use is also produced by hydration of ethylene byproduct in the petroleum industry.

5.1.2 Ethanol Fuel

When compared with gasoline, ethanol has a higher octane number, broader flammability limits, higher flame speeds, and higher heats of vaporization. These properties allow for a higher compression ratio, shorter burn time, and leaner burn engine, resulting in a higher efficiency. Ethanol is an oxygenated fuel that contains 35% oxygen, which reduces particulate and nitrogen oxides emissions from combustion. Disadvantages of ethanol include a lower energy density than gasoline, corrosiveness, difficult cold start due to low vapor pressure, low flame luminosity, miscibility with water, and some toxicity to the ecosystems.

Due to its high octane number, ethanol is appropriate as a gasoline-mixed fuel for gasoline engines. But ethanol is not suitable for diesel engines, as its low cetane number and high heat of vaporization impede self-ignition. In Brazil, pure ethanol or a blend with gasoline (gasohol: 24% ethanol + 76% gasoline) is used. In some parts of the United States, 10% ethanol is added to gasoline, known as E10 fuel. Blends with higher ethanol content can also be used by flex fuel vehicles (FFV), which can operate with up to 85% ethanol (E85) added. Additional countries using ethanol–gasoline blends include Canada (E10 and E85 for FFV), Sweden (E5 and E85 for FFV), India (E5), Australia (E10), Thailand (E10), China (E10), Columbia (E10), Peru (E10), and Paraguay (E7) (Kadiman, 2005).

5.2 Ethanol in Ancient Times

Ethanol has been used by humans since ancient times in a variety of ways, including as meals, as medicine, as a relaxant, as an aphrodisiac, for euphoric effects, for recreational purposes, for artistic inspiration, and in religious ceremonies. In ancient Egypt, alcohol was produced by naturally fermenting vegetative materials, which provided only a dilute concentration. To increase the concentration, the Chinese discovered distillation, as evident from the dried residues on ancient pottery in China. Based on the residue found in pottery jars from the Neolithic village of Jiahu (Henan province), a mixed fermented beverage of rice, honey, and fruits was being produced as early as 9,000 years ago. At about the same time, wines from barley and grapes were being made in the Middle East, as shown in the recipes engraved on clay tablets and art in Mesopotamia.

The Hindu Ayurvedic texts describe both the beneficial medical uses of alcoholic beverages and the consequences of intoxication and alcoholic diseases. As a result, most current Eastern religions (e.g., Hinduism, Buddhism, and Islam) prohibit the consumption of alcohol. On the other hand, in some Western religions, alcohol consumption has some symbolic and religious significance, for example, in Christian Eucharists, on the Jewish Shabbat and Passover, in the Dionysus ritual of Greco-Roman religion, and so on. During the Middle Ages (fifth to sixteenth centuries) in Europe, beer was consumed by the whole family, including children. Documents also show nuns having a daily allowance of ale.

South Americans produced alcohol from cassava and maize (cauim, chichi), which had to be chewed to convert starch into sugars for fermentation. In ancient Japan, this chewing technique was also used to convert rice and other starchy crops. Although distillation techniques date back to ancient China, Persian alchemist Zakariya Razi isolated ethanol in a relatively pure form. In fact, the word "alcohol" is derived from the Arabic word for "finely divided," a reference to distillation. In 1796, Johann Tobias Lowitz obtained pure ethanol by filtering distilled alcohol through activated charcoal. Ethanol as a compound of carbon, hydrogen, and oxygen elements was described by Antoine Lavoisier, and its chemical formula was determined by Nicolas-Theodore de Saussure in 1808. Five decades later, Archibald Scott Couper published the structural formula of ethanol.

Ethanol can also be produced synthetically from ethylene, a byproduct from the petroleum industry. The original process was developed by Michael Faraday in 1828 in which ethanol was prepared by acid-catalyzed hydration of ethylene; in fact, a similar process is used today for industrial level synthesis.

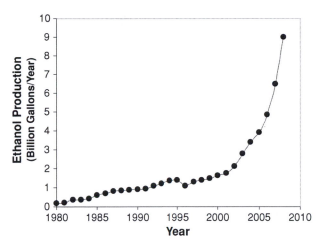

Figure 5.1. Ethanol productions in the United States.

As early as 1840, ethanol was used as a lamp fuel in the United States. Later, a tax levied during the Civil War made this use uneconomical, but the tax was repealed in 1906 so that the Ford Model T automobile could use ethanol as a fuel. During the Prohibition in the 1920s, use of ethanol as fuel was discouraged due to its possible misuse in beverages.

5.3 Current Ethanol Production

During the OPEC oil embargo of the 1970s, ethanol fuel reemerged as a fuel extender during the gasoline shortage. Later, ethanol was used as a replacement for lead, as a cleaner burning octane enhancer. In 1995, about 93% of the ethanol in the world was produced by the fermentation method and about 7% by the synthetic method. As a result of energy security needs, farmer incentives, and clean air regulations, U.S. ethanol demand has grown from 175 million gallons in 1980 to about 9,000 million gallons in 2008 (Figure 5.1).

The global ethanol production has increased from 10.7 billion gallons in 2004 to 20.4 billion gallons in 2008. U.S. ethanol production has exceeded that of Brazil (Table 5.1).

Ethanol demand is expected to more than double in the next ten years. At present, ethanol accounts for about 94% of the global biofuel production, with the majority being produced from sugarcane and corn. Brazil and the United States are the world leaders in ethanol production; together, they account for 78% of the world's ethanol production. Ethanol is a well-established biofuel used for transport and in the industry sectors in several countries, notably in Brazil. Brazil has been using ethanol as a fuel since 1925; at that time, the production of ethanol was 70 times more than the production and consumption of petrol. Currently, the petrol sold in Brazil contains 25% ethanol. The United States has used ethanol produced from corn since 1980, and the current petrol contains 10% ethanol (Demirbas and Balat, 2006).

Table 5.1. *Global ethanol production, in billion gallons per year (Agranet, 2009)*

Country	Year 2004	Year 2008	Share year 2008 (%)
United States	3.40	8.93	43.8
Brazil	3.87	6.90	33.9
China	0.92	1.02	5.0
India	0.32	0.61	3.0
France	0.22	0.40	1.9
Canada	0.06	0.26	1.3
Germany	0.06	0.22	1.1
Thailand	0.65	0.15	0.7
Russia	0.20	0.15	0.7
Spain	0.09	0.13	0.6
South Africa	0.10	0.11	0.5
United Kingdom	0.08	0.11	0.5
Remaining countries	1.35	1.40	6.9
World	10.75	20.37	100.0

The fermentation process can use any sugar-containing material to produce ethanol. The sugar-containing agricultural raw materials can be classified into three categories: (1) sugar, (2) starch, and (3) cellulose. Sugars (e.g., from sugarcanes, molasses, sugar beats, and fruits) can be directly fermented using yeast to produce ethanol. Starch (e.g., from corn and other grains, potatoes, root crops) and cellulose (e.g., from wood, agricultural residue, grasses) are first converted to sugar by hydrolysis/pretreatment and then fermented. Due to the sturdy nature of cellulose molecules and biomass structure, hydrolysis to sugar is a difficult process. Cellulosic ethanol is discussed in Chapter 6.

Most of the ethanol in the United States is produced from corn, and ethanol in Brazil is produced from sugarcane. The ethanol yield in the United States and Brazil is 400 and 870 gallons/acre/year, respectively. European countries use beet molasses to produce ethanol, as it yields substantially more ethanol per acre than wheat. The advantages of sugar beet include lower cycle of crop production, higher yield, high tolerance of a wide range of climatic variations, and low water and fertilizer requirements.

5.4 Fermentation

Fermentation of sucrose is performed using commercial yeasts such as *Saccharomyces cerevisiae*. First, the invertase enzyme contained in the yeast converts sucrose to glucose and fructose, as

$$C_{12}H_{22}O_{11} \rightarrow C_6H_{12}O_6 + C_6H_{12}O_6 \qquad (5.1)$$
$$\text{Sucrose} \qquad \text{Glucose} \qquad \text{Fructose}$$

Second, zymase, another enzyme also present in yeast, converts glucose and fructose into ethanol and carbon dioxide (CO_2), as

$$C_6H_{12}O_6 \rightarrow 2C_2H_5OH + 2CO_2 \qquad (5.2)$$

Basically, glucose and fructose are used by yeast to produce cellular energy, and ethanol and CO_2 are produced as metabolic waste products. Because the process does not require oxygen, the fermentation is carried out in the absence of oxygen, and the process is termed anaerobic. Ethanol fermentation is responsible for the rising of bread dough, the production of ethanol in alcoholic beverages, and much of the production of ethanol fuel.

An important step in the production of ethanol from sugar is the fermentation process and the microorganism involved. Key performance parameters include temperature range, pH range, alcohol tolerance, osmotic tolerance, inhibitor tolerance, growth rate, productivity, specificity, yield, and genetic stability. Significant efforts have been invested in finding the optimum microorganism and fermentation conditions. A microorganism of choice is *S. cerevisiae*, also known as baker's yeast, which has been used in production of food and beverages for several millennia. *S. cerevisiae* is an efficient ethanol producer from glucose (a six-carbon sugar), does not require oxygenation, requires low pH, and has a relatively high tolerance to ethanol and other inhibitors.

Traditionally, fermentation is carried out in batch reactors; hence, the microorganism is exposed to a high concentration of ethanol (an inhibitor for fermentation) toward the end of the process. Therefore, microorganisms with high ethanol tolerance are needed that can use low-cost fermentation substrates. A promising candidate, *Zymomonas mobilis* (a Gram-negative bacterium), is emerging as an alternative organism for large-scale fuel ethanol production. Advantages of *Z. mobilis* include high sugar uptake and ethanol yield, high ethanol tolerance, low biomass production, does not require controlled oxygenation, and amenability to genetic enhancements (Gunasekaran and Raj, 1999). A disadvantage is that *Z. mobilis* is limited to the fermentation of D-glucose, D-fructose, and sucrose. Due to its specific substrate spectrum as well as the undesirability of its biomass to be used as animal feed, this species cannot readily replace *S. cerevisiae* in ethanol production (Bai, Anderson, and Moo-Young, 2008).

Additional innovations are coming in the form of immobilization of the microorganism and continuous removal of ethanol. Immobilization of cells can potentially increase the cell mass concentration in the fermenter to increase the process productivity and minimize the production costs, while offering advantages over free cell fermentation operations (Santos et al., 2008). There has been a surge in attempts to find a suitable cell carrier material, for example, loofa sponge, petiolar felt-sheath of palm, and sugarcane bagasse. The advantages accruable from such biomaterials are reusability, freedom from toxicity problems, mechanical strength for necessary support, and opening of spaces within the matrix for growing cells (Chandel et al., 2009). However, being the primary metabolite, ethanol production is tightly coupled with the growth of yeast cells, indicating that yeast must be produced as a coproduct (Figure 5.2). Glucose fermentation using *S. cerevisiae* proceeds with glycolysis, through which one molecule of glucose is metabolized and two molecules of pyruvate are produced. Pyruvate, under anaerobic conditions, reduces to ethanol and CO_2. On glucose mass basis, theoretical yields of ethanol and CO_2 are 51.1%

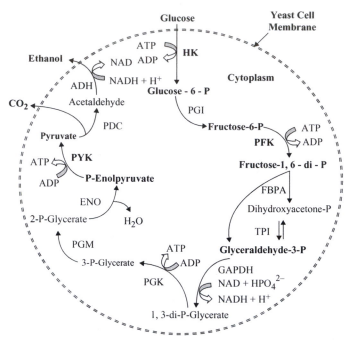

Figure 5.2. Metabolic pathway of ethanol fermentation in *S. cerevisiae* (Bai et al., 2008; reproduced with permission from Elsevier). Abbreviations: HK, hexokinase; PGI, phosphoglucoisomerase; PFK, phosphofructokinase; FBPA, fructose bisphosphate aldolase; TPI, triose phosphate isomerase; GAPDH, glyceraldehyde 3-phosphate dehydrogenase; PGK, phosphoglycerate kinase; PGM, phosphoglyceromutase; ENO, enolase; PYK, pyruvate kinase; PDC, pyruvate decarboxylase; ADH, alcohol dehydrogenase.

and 48.9%, respectively (Bai et al., 2008). During glycolysis, two adenosine triphosphates (ATPs) are produced, which are used for energy needs of yeast cell synthesis. The use of ATP is necessary in order to continue the fermentation without interruption. Hence, the ethanol production is tightly coupled with the growth of yeast cells (Bai et al., 2008). An accumulation of ATP can inhibit phosphofructokinase, one of the most important regulation enzymes in glycolysis. Immobilization of yeast cells restrains their growth and makes its removal from the system more difficult. Hence, immobilized yeast cell technology has not been adopted. On the other hand, the self-immobilization of yeast cells through their flocculation can effectively overcome some of these drawbacks.

Both *S. cerevisiae* and *Z. mobilis* are excellent ethanol producers from glucose and sucrose but are incapable of fermenting pentose (five-carbon) sugars. Separate xylose-fermenting microorganisms are found among bacteria, yeast, and filamentous fungi. Efforts are underway to introduce into *Saccharomyces* a pathway for xylose metabolism from a natural xylose-using organism, for example, genes encoding for xylose reductase and xylitol dehydrogenase enzymes, which are present in the natural xylose-using yeast *Pichia stipitis*. However, only low amounts of ethanol have been obtained from such efforts (Martin et al., 2002).

5.5 Ethanol from Sugarcane

After harvesting, the sugarcanes are transported to the plant, where washing, chopping, and shredding are performed. Prepared sugarcanes are then fed into a juice extractor, which produces liquid containing 10–15% sucrose and solid residue (bagasse). The extractor designs have evolved to extract maximum sucrose from the sugarcanes. The juice is filtered, treated with chemicals, and pasteurized. Then, the juice is further filtered to produce vinasse, a fluid rich in sucrose. Water from vinasse is removed by evaporation to produce syrup. The sucrose is precipitated from the syrup by a crystallization process, resulting in slurry-containing clear crystal and molasses. A centrifuge is used to separate the crystals from molasses. Crystals are washed with the addition of steam and then dried by air flow. Molasses is then sent for ethanol fermentation. Bagasse, the solid residue after extraction of the sugarcane juice, is mostly used for producing steam and electricity required for the cane-processing plant.

Blackstrap molasses contains sucrose (35–40 wt%), invert sugars such as glucose and fructose (15–20 wt%), and nonsugar solids (28–35 wt%). In sugar manufacturing, blackstrap (syrup) is collected as a byproduct. The typical process includes the following steps: (1) molasses is diluted to a mash containing about 10–20 wt% sugar; (2) pH of the mash is adjusted to about 4–5 by adding a mineral acid; (3) yeast is added to the mash and fermentation is carried out nonaseptically at 20–32°C for about 1–3 days; and (4) the fermented beer, which typically contains about 6–10 wt% ethanol, is then taken for distillation to recover ethanol.

5.6 Ethanol from Corn

All potable alcohol and most fermented industrial alcohol in the United States are currently made from grains. Typically, corn contains 15% moisture, 7.7% protein, 3.3% oil, and 62% starch. Fermentation of starch from corn is somewhat more complex than fermentation of sugars because starch must first be converted to sugar by hydrolysis, which is typically carried out enzymatically by diastase present in sprouting grain or by fungal amylase. The resulting dextrose is fermented to ethanol with the aid of yeast producing. A second coproduct of unfermented starch, fiber, protein, and ash known as distillers grain (a high-protein cattle feed) is also produced.

The main component of interest in ethanol production is the starch content of corn. Starch is a polymer of glucose subunits (and $C_{12}H_{16}O_5$ repeat unit) linked via α-1,4 linkages, with some branches formed by α-1,6 linkages (Figure 5.3). The starch polymer is highly amorphous, making it readily digested by human and animal enzyme systems for hydrolysis into glucose units.

There are two distinct methods for processing corn, wet milling and dry milling, and each method generates unique coproducts.

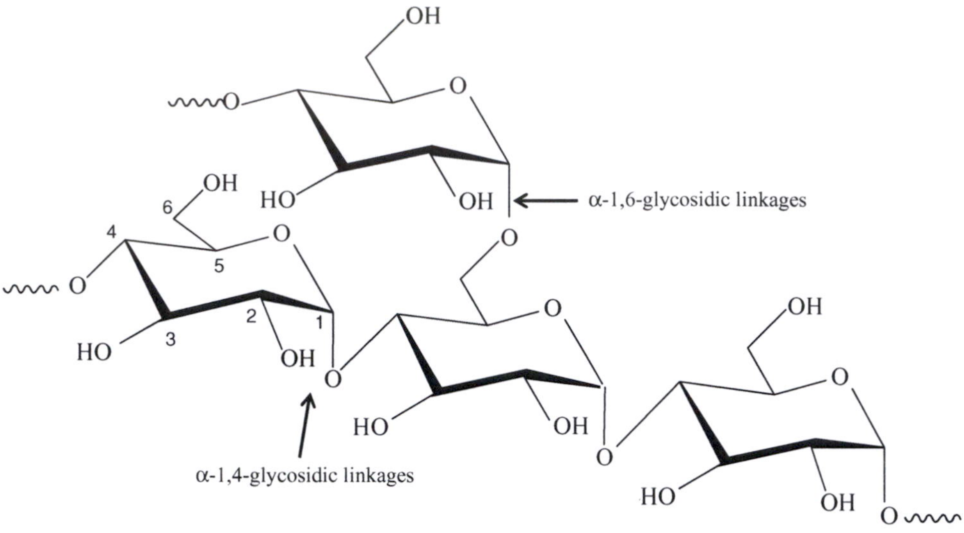

Figure 5.3. The polymeric structure of glucose in starch.

5.6.1 The Corn Wet-Milling Process

In wet milling (or wet grind), corn is cleaned and soaked for about 40 hours at 50°C in water containing diluted sulfur dioxide ($\sim$550 ppm). During the soaking process, soluble nutrients (proteins, vitamins, minerals) leach out in water; this water is later concentrated to produce condensed corn fermented extractives (corn steep water or liquor), which is a high-energy liquid feed ingredient. Then, corn germ is removed from the soaked kernels, germs are further processed to recover corn oil, and remaining germ meal is used as feed. Now the corn kernels are screened to remove bran, leaving the starch and gluten protein to pass through the screen. The gluten protein is separated by centrifugation, causing lighter protein to float at the top and heavier starch to settle at the bottom; the gluten meal is used as a high-protein feed. The remaining starch is taken for ethanol fermentation. The coproducts from this process account for the 25–30% of the corn processed (Davis, 2001).

5.6.2 The Corn Dry-Milling Process

In dry milling (or dry grind), corn is cleaned of foreign materials and hammer milled to produce cornmeal. The cornmeal is liquefied by mixing with water at 5–6 pH and 80–90°C, and α-amylase enzyme is added to facilitate the hydrolysis of cornstarch to dextrin, a long-chain sugar. After liquefaction, the mash is cooked to release amylopectin contained in starch and to kill unwanted lactic acid-producing bacteria. Then, the mash is cooled, and glucoamylase enzyme is added to convert dextrin into simple-sugar dextrose. The simple sugars are then fermented using yeast (*S. cerevisiae*) to produce ethanol and CO_2. The fermentation is completed in 40–60 hours, and the resulting beer is sent for distillation. All solids (protein, fat, and fiber) and

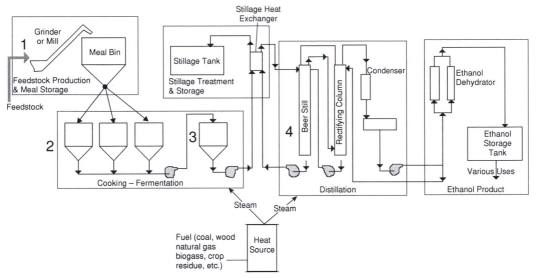

Figure 5.4. Flowsheet for a typical corn ethanol plant (DNR, Louisiana, 2009).

water are collected from the distillation base and referred to as whole stillage, which is then centrifuged to separate coarse solids from the liquid. The liquid (also known as thin stillage) is recycled to the liquefaction process or concentrated to produce corn condensed distillers solubles. The coarse solids collected from the centrifuge and condensed solubles are then combined and dried in a rotary dryer to form the feed coproduct corn distillers dried grains with solubles (Davis, 2001). Figure 5.4 shows the flow sheet for the production of bioethanol from cereal grain.

Much of U.S. ethanol production is via the dry-milling process. The key difference between the two processes is the removal of nonstarch components. In dry-milling, these are removed after fermentation, whereas, in the wet-milling process, these components are removed before fermentation. The wet-milling process has more steps but can produce valuable coproducts. On the other hand, dry-milling process has a lower cost.

5.6.3 Byproducts

Corn-based ethanol produces a variety of byproducts that have a high utility. Corn oil is supplied as vegetable oil or upon esterification as biodiesel. Corn gluten, largely produced in the wet-milling process, is a popular cattle protein feed containing intermediate protein products that are rich in digestible fibers. Corn gluten meal, a golden-yellow high-protein portion of the corn kernel, is used primarily in the swine and poultry industries due to high content of yellow pigment in addition to its nutritional values. Condensed steepwater solubles are an excellent source of soluble protein for liquid beef supplements. The wet distillers grains and distillers grains with solubles contain nutrients that are concentrated approximately three times more than those found in corn. Dried distillers grains and distillers grains with solubles are marketed widely around the world as a feed commodity. Like corn gluten meal, dried distillers grains are a good protein source for cattle. On the other hand, the dry-milling ethanol industry is now developing the ability to fractionate the corn kernel, so that germ can be used for extraction of oil and high-protein distillers grain can be obtained, which are more suitable for the swine and poultry industries.

5.7 Separation and Purification

Fermentation only yields alcohol concentrations of about 10%, whereas the fuel application requires 100% pure ethanol. Hence, significant effort and energy are spent on the removal of water. In the first step, a distillation or beer column is used; a large amount of water leaves with the solids, and product containing 37% ethanol is obtained. This product is then concentrated in a rectifying column to a concentration just below the azeotrope (95%). (Note: Simple distillation cannot increase the concentration past the azeotrope concentration.) The remaining ethanol-lean bottoms product is fed to the stripping column to remove additional water. The distillate from the stripping is combined with the feed to the rectifier column.

To further remove water from the 95 mol% ethanol, several techniques are available, including (1) lime or salt drying, (2) molecular sieve drying, (3) azeotropic distillation, and (4) pressure-swing distillation. Out of these, molecular sieve drying is the most popular.

In drying using lime, calcium oxide (lime) is mixed with the ethanol/water mixture. Lime, upon reacting with water, forms calcium hydroxide, which can be separated; thus, the water is removed. Calcium hydroxide can be regenerated to lime by heating.

In molecular sieve drying, 3-angstrom pore-size zeolite molecular sieves (solid pellets of about 3 mm diameter) are used. The pores are large enough for water molecules (about 2.8 angstrom in size) to penetrate but too small for ethanol molecules (4.4 angstrom in size). The water adsorption capacity is very large, as zeolite molecular sieves can adsorb up to 22 wt% water. The water-soaked zeolite is simply filtered out from the pure ethanol and then regenerated by heating. In addition to the molecular sieves, a variety of locally available plant materials (e.g., cornmeal, straw, and sawdust) can be used to make water or ethanol adsorbents.

In azeotropic distillation, a small amount of benzene is added to make a ternary azeotrope, which is more volatile than the ethanol–water azeotrope. The ternary azeotrope is distilled out of the ethanol–water mixture, extracting essentially all of the water in the process. The bottom product is water-free ethanol with several ppm of residual benzene. In later designs, benzene, due to its human toxicity, has been replaced by cyclohexane.

In the pressure-swing distillation, a pressure less than atmospheric is used in distillation, which causes the ethanol–water azeotrope to shift to a more ethanol-rich mixture. Hence, distillation first yields an ethanol–water mixture containing more than 96% ethanol. The mixture is again distilled at atmospheric pressure to obtain pure ethanol.

5.8 Summary

Ethanol from carbohydrates by fermentation is a historical industry with very early application in beverage making. The recent use of ethanol as fuel has increased its production. Most ethanol is currently being produced from sugar cane or from corn. Yeast is used to ferment sugars into ethanol. In the case of carbohydrates (such as corn), a pretreatment step of converting carbohydrate into sugars is needed. Currently, the corn ethanol industry uses either a dry-milling or a wet-milling process. Upon fermentation, ethanol content is only about 10%, which requires a significant effort in separation to produce the pure ethanol needed for fuel use. Distillation can concentrate ethanol to just below the azeotropic concentration (95 mol%), after that, specialized separations (e.g., molecular sieve, azeotropic distillation, lime drying) are needed. Efforts are underway to improve the microorganism and separation processes to reduce the overall cost of ethanol production. With the current trend in ethanol use, demand is likely to increase significantly in the near future.

6 Ethanol from Biomass by Fermentation

6.1 Challenges with Corn-Based Ethanol

Today, the world's ethanol supply is mainly derived from U.S. corn or Brazilian sugarcane. Corn and other grain (as farmers planted corn instead of other grains) prices have soared internationally, and the corn-to-ethanol industry has been blamed for driving up food prices worldwide (Figure 6.1). The Food and Agriculture Organization of the United Nations and the World Bank have said that soaring world food prices are due in part to increased demand for biofuels. However, additional factors that contribute to increased food prices include higher energy prices and increased consumption of meat and dairy products in developing economies of China and India. Note that it requires about 13 kg of corn to raise 1 kg of meat; hence, a shift toward nonvegetarian diet will put even more pressure on the food supply. The U.S. energy bill of 2007 allows for the doubling of corn-based ethanol with a ceiling of 15 billion gallons per year. The bill also calls for an additional 21 billion gallons per year of advanced biofuels, including 16 billion gallons per year of ethanol from nonfood sources, by 2022. The energy policy means that corn is likely to rule the U.S. ethanol industry for many years. However, soaring food prices and questions about whether corn-for-fuel can reduce global warming have sparked a debate about whether the United States is going down the wrong road in the search for alternatives to fossil fuels. The policy is expected to put additional pressure on the food supply if nonfood-based biofuels are not developed.

Although corn kernel is more easily fermented into ethanol than cellulosic biomass, corn kernel is also about threefold more expensive than biomass on a per-unit energy or mass basis. Additionally, the theoretical amount of ethanol produced per acre of land via corn kernel is much lower than that from biomass. Relative to corn kernel, production of a perennial cellulosic biomass crop, such as switchgrass, requires lower amounts of energy, fertilizer, pesticide, and herbicide and is accompanied by less erosion and improved soil fertility.

6.2 Cellulose in Biomass

Biomass is primarily composed of cellulose, hemicelluloses, and lignin, along with small amounts of protein, pectin, extractives (e.g., sugars, chlorophyll, waxes), and

Table 6.1. *Cellulose, hemicellulose, and lignin content in biomass (Kumar et al., 2009)*

Biomass	Cellulose (wt%)	Hemicellulose (wt%)	Lignin (wt%)
Coastal bermudagrass	25	35.7	6.4
Corn cobs	45	35	15
Cotton seed hairs	80–95	5–20	0
Grasses	25–40	35–50	10–30
Hardwood stems	40–55	24–40	18–25
Leaves	15–20	80–85	0
Newspaper	40–55	25–40	18–30
Nutshells	25–30	25–30	30–40
Paper	85–99	0	0–15
Primary wastewater solids	8–15		
Softwood stems	45–50	25–35	25–35
Solid cattle manure	1.6–4.7	1.4–3.3	2.7–5.7
Sorted refuse	60	20	20
Swine waste	6.0	28	NA[a]
Switchgrass	45	31.4	12
Waste papers from chemical pulps	60–70	10–20	5–10
Wheat straw	30	50	15

[a] NA, not applicable.

ash. Two large carbohydrate categories that have significant value are cellulose and hemicelluloses; when combined, they are referred to as holocellulose. The lignin fraction consists of nonsugar-type molecules. The composition can vary depending on the source of the biomass, as shown in Table 6.1. Even in a single biomass source, variations arise due to age, state of growth, and other conditions.

Cellulose is located primarily in the secondary cell wall and is part of the organized fibrous structure. Cellulose is a remarkable pure organic polymer, consisting solely of units of anhydroglucose held together in a giant straight-chain molecule (Demirbas, 2000a). These anhydroglucose units are bound together by β-(1,4)-glycosidic linkages. Due to this linkage, cellobiose is established as the repeat unit for cellulose chains (Figure 6.2). Cellulose must be hydrolyzed to glucose ($C_6H_{12}O_6$) before fermentation to ethanol.

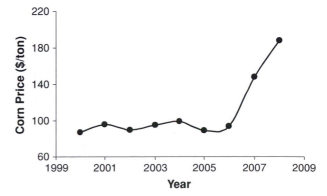

Figure 6.1. U.S. corn prices.

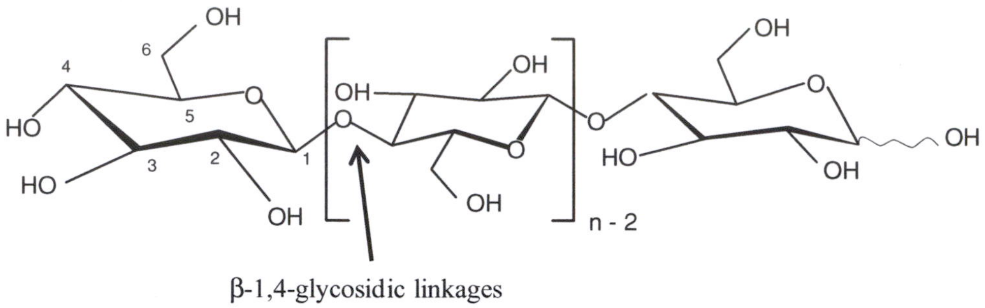

β-1,4-glycosidic linkages

Figure 6.2. Structure of cellulose polymer with *n-2* anhydroglucose subunits connected by β-(1,4)-glycosidic linkages. The left end monomer is nonreducing and the right end monomer is reducing-type glucose.

The number of glucose units in a single cellulose molecule can vary from 500 to 5,000, depending on the source of biomass. In a given fiber, cellulose molecules are linked by intramolecular and intermolecular hydrogen bonding. The chains are arranged in parallel, forming a crystalline supermolecular structure. Then, bundles of linear cellulose chains (in the longitudinal direction) form a microfibril (Figure 6.3), which is oriented in the cell wall structure (Fratzl, 2003; Hashem et al., 2007). Single hydrogen bonds are weak, but an organization of thousands of hydrogen bonds imparts a strong bonding among cellulose molecules. This is why cellulose

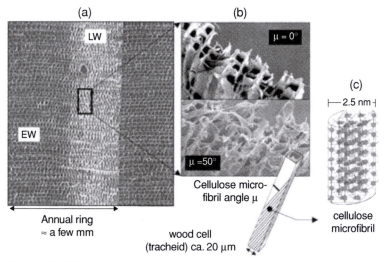

Figure 6.3. Hierarchical structure of spruce wood. (a) Cross-section through the stem showing earlywood (EW) and latewood (LW) within an annual ring. Latewood is denser than earlywood because the cell walls are thicker. The breadth of the annual rings varies widely depending on climatic conditions during each particular year. (b) Scanning electron microscopic pictures of fracture surfaces of spruce wood with two different microfibril angles. One of the wood cells (tracheids) is drawn schematically, showing the definition of the microfibril angle between the spiralling cellulose fibrils and the tracheid axis. (c) Sketch of (the crystalline part of) a cellulose microfibril in spruce (Fratzl, 2003; reproduced with permission of Elsevier).

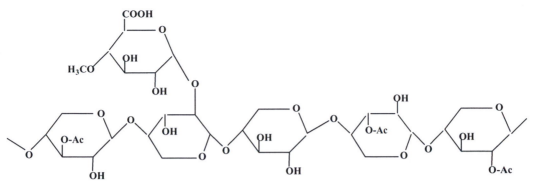

Figure 6.4. A segment of hardwood xylan, *O-acetyl-4-O-methylglucuronoxylan* (adapted from Bobleter, 1994).

cannot be easily broken down. Cellulose is insoluble in most solvents and has a low accessibility to acid and enzymatic hydrolysis. However, some animals (e.g., cows and termites) have intestinal microorganisms that can break down cellulose into sugars using β-glycosidase enzymes. Biomass may also contain a very small portion of amorphous cellulose, which is more susceptible to enzymatic hydrolysis.

In contrast to cellulose, which is a remarkable pure polymer, hemicelluloses contain varying monosachharide repeat units. Hemicelluloses are polymers of pentoses (xylose, rhamnose, and arabinose), hexoses (glucose, mannose, and galactose), and uronic acids (D-glucuronic, D-galactouronic, and 4-O-methylglucuronic acids, etc.) repeat units. Another difference from cellulose is that hemicellulose molecules have branch chains, although the backbone chain can be a homopolymer (generally consisting of a single sugar repeat unit) or a heteropolymer (mixture of different sugars). The main function of hemicelluloses is to act as the cement material holding together the cellulose micelles and fibers (Theander, 1985). Hemicelluloses are much easier to hydrolyze when compared with cellulose, do not aggregate even when they cocrystallize with cellulose chains, and are largely soluble in alkali solutions (Timell, 1967; Wenzl, 1970; Goldstein, 1981).

Mainly, the hemicelluloses in deciduous woods are made of pentosans and in coniferous woods are made of hexosanes. Among the most important sugars of the hemicellulose component is xylose ($C_5H_{10}O_5$), which makes the xylan backbone. The xylan chain consists of xylose units that are linked by β-(1,4)-glycosidic bonds and branched by α-(1,2)-glycosidic bonds with 4-O-methylglucuronic acid groups (Figure 6.4) (Hashem et al., 2007). The molecular structure can vary from the type of biomass.

Lignin is a highly cross-linked polymer of phenolic monomers (e.g., guaiacyl propanol, p-hydroxyphenyl propanol, and syringyl alcohol). In general, softwoods have high lignin content and grasses have very low lignin content. The main function of lignin in plants is to impart structural support and to provide impermeability and resistance against microbial attack. Hence, lignin provides resistance in the use of cellulose and hemicelluloses for ethanol production.

Table 6.2. *Relative abundance of individual sugars in carbohydrate fraction of wood (Goldstein, 1981)*

Sugar	In softwoods (wt%)	In hardwoods (wt%)
Glucose	61–65	55–73
Xylose	9–13	20–39
Mannose	7–16	0.4–4
Galactose	6–17	1–4
Arabinose	<3.5	<1

6.3 Sugars from Cellulose

The key technology step in the production of ethanol from cellulose (and hemicelluloses) is the hydrolysis of cellulose to sugar. After this, the established sugar-to-ethanol fermentation technology can be used (as described in Chapter 5). Both cellulose and hemicelluloses portions of biomass provide sugars. In addition to glucose (six-carbon sugar), hemicelluloses also provide pentose (five-carbon sugar). The relative abundance of individual sugars in the carbohydrate fraction of wood is shown in Table 6.2.

Because pentose molecules (five-carbon sugars) comprise a good percentage of available sugars, the ability to recover and ferment them into ethanol is important for the efficiency and economics of the process. For example, the hydrolysate from corn stover contains about 30% of the total fermentable sugars as xylose (Kumar, et al., 2009). Six-carbon sugars (e.g., hexoses, glucose, galactose, and mannose) are readily fermented to ethanol by organisms such as baker's yeast (*Saccharomyces cerevisiae*) used in the brewing industry. Recently, special microorganisms have been genetically engineered that can ferment five-carbon sugars into ethanol with relatively high efficiency (Becker and Boles, 2003; Jeffries and Jin, 2004; , Karhumaa et al., 2006; Ohgren et al., 2006). Bacteria (e.g., *Zymomonas mobilis*) have drawn special attention from researchers because of their speed of fermentation.

6.4 Factors Affecting Lignocellulose Digestibility

Lignocellulosic biomass typically contains 55–75% cellulose plus hemicelluloses by weight. The key parameter in ethanol production is the digestibility of biomass to produce sugars. Various factors that affect the biomass digestibility are discussed below.

6.4.1 Lignin Content

Most of the lignin is concentrated between the outer layers of the cellulose/hemicelluloses fibers, providing a structural rigidity. Some of the lignin is intertwined among cellulose and hemicelluloses. In addition, lignin provides protection against the microbial degradation of the fibers by the inhibitory effect of the phenols. Hence, lignin content and type significantly affect the hydrolysis of biomass. Lignin can also bind with cellulase enzyme, resulting in less availability of the enzyme for hydrolysis. It has been shown that the chemical and physical structures of lignin play a significant role in determining the magnitude of inhibition it contributes to hydrolysis, and the structure of lignin is heavily dependent on the conditions of the substrate pretreatment (Chandra et al., 2007). Phenols, phenol aldehydes, and phenol ketones (Hibbert's ketones) are potent inhibitors; hence, their formation during the pretreatment process should be minimized.

6.4.2 Hemicelluloses Content

Hemicelluloses removal can significantly improve the hydrolysis of cellulose. In addition, the hemicelluloses hydrolysis itself can improve the sugar yield available for ethanol production. Hence, pretreatment conditions are optimized to remove lignin and hemicelluloses while maximizing sugar yield. The hemicelluloses degradation products furfural and hydroxymethylfurfural inhibit subsequent fermentation. Furfural and 5-hydroxymethylfurfural are the result of degradation of pentoses and hexoses, respectively. Furfural may react further to yield formic acid or it may polymerize. With process optimization it is possible to decrease the inhibitory compounds and increase total sugar concentrations. Ultimately, it is the hydrolysis of solid cellulose that provides the majority of the glucose for ethanol production. Increasing the total recovery of sugars while minimizing inhibitor production must be weighed against the negative effect of the substrate hemicellulose (Chandra et al., 2007).

6.4.3 Acetyl and Other Inhibitor Content

In hemicelluloses, about 70% of xylan is randomly acetylated (i.e., has CH_3COO- groups), which can hinder the enzymatic action. It has been shown that removing acetyl groups can enhance biomass digestibility through increased swellability (Zhu et al., 2008). However, the impact is not as pronounced as that of lignin content removal. Techniques such as potassium hydroxide deacetylation can be used to reduce the acetyl content.

Overall, degradation products such as phenols, furans, and carboxylic acids are fermentation inhibitors. For economically feasible ethanol production, either removal of such compounds or detoxification is needed. For example, the recirculation of the stillage water in the process will minimize water consumption, but detoxification prior to recirculation is required unless the inhibitory compounds are removed. However, the cost of detoxification should be carefully weighed against the benefit of improvement in the fermentability. Furthermore, separating the liquid hemicellulose fraction containing the inhibiting compounds from the cellulose fraction prior to enzymatic hydrolysis and fermentation can reduce the content of inhibitory compounds (Klinke, Thomsen, and Ahring, 2004).

6.4.4 Cellulose Crystallinity and Degree of Polymerization

Similar to starch, cellulose is a polymer of glucose. The monomer arrangement in starch is such that the polymer molecules do not assemble into a crystalline structure; hence, starch remains amorphous and readily accessible to enzymatic hydrolysis and digestion. On the other hand, the glucose monomers in cellulose are arranged in such a way that the polymer molecules can arrange into a crystalline structure (crystalline structures, if possible, are preferred by nature over the amorphous structures). Cellulose crystallinity provides a protection against enzymatic attack and solubilization in water; hence, cellulose crystallinity directly affects digestibility. While measuring crystallinity, one needs to separate the crystalline portion of cellulose from that of the whole biomass, as lignin is amorphous and its removal will show up as an increase in the overall crystallinity of the biomass when the cellulose crystallinity is constant.

It is difficult to assess solely the effect of degree of polymerization (DP) on digestibility, as the DP is closely linked with the crystallinity and the accessible surface area. If pretreatment cleaves the internal cellulose bonds (for example, in acidic pretreatment), then it becomes easier for enzymes (exocellulases) to attack the remaining lower DP polymer chains. This is in contrast with pretreatment that protects internal cellulose bonds (for example, alkaline pretreatment), causing not much decrease in DP.

6.4.5 Surface Area of Pore Volume

The cellulase action on cellulose is in the liquid–solid phase; hence, the action is a surface reaction. A high surface area and enough pore space for enzyme to reach cellulose enhance the digestibility. Lignocellulosic biomass such as wood possesses limited surface area available for cellulase to work on. Hence, increase of surface area and pore volume is of utmost importance in pretreatment. For example, in the case of pulp fibers, the rate and extent of hydrolysis have been directly correlated with the initial specific surface area of the fibers. The surface area of pulp fibers is composed of exterior surface area due to fiber length and width, and interior surface area is governed by the size of the lumen and the number of fiber pores and cracks. Both types of surface area are important in biomass conversion (Figure 6.5). Particle size is inversely related to the specific exterior surface area. Therefore, biomass with smaller particle size would be expected to hydrolyze at a faster rate. In fact, a direct correlation has been found between initial pore volume and interior surface area of biomass and their extent of hydrolysis. It has been proposed that the efficacy of cellulose hydrolysis is enhanced when the pores of the substrate are large enough to accommodate both large and small enzyme components to maintain the synergistic action of the cellulase enzyme system (Galbe and Zacchi, 2007).

6.5 Biomass Pretreatment

Due to the robust molecular structure of biomass, pretreatment is required before the enzymatic hydrolysis step to produce fermentable sugars economically (Laxman

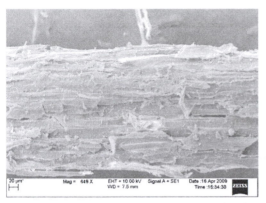 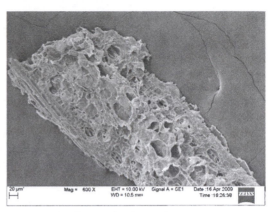

Figure 6.5. Original switchgrass particle (left) compared with after hydrothermal pretreatment (right) with increased surface area and pore volume (Kumar et al., 2009).

and Lachke, 2008). Pretreatment increases the enzymatic access in several ways, including (1) removing hemicelluloses, (2) removing acetyl groups from the hemicelluloses, (3) removing or altering lignin, (4) expanding the structure to increase pore volume and internal surface area, (5) reducing the degree of polymerization in cellulose, and (6) decrystallizing cellulose. A single pretreatment cannot achieve all of these things, and multiple pretreatments or one severe pretreatment is likely to be very expensive. Hence, a process optimization scheme is used to select a pretreatment scheme that minimizes the overall ethanol cost. Over the years, a number of different technologies have been developed for pretreatment of lignocellulose (Table 6.3). Different pretreatment processes affect biomass in different ways. Desirable characteristics of a pretreatment process include (1) requires minimal energy and chemical input, (2) is suitable for a variety of biomass feedstocks, (3) requires minimal capital and operating expenses, (4) is scalable to the industrial scale, and (5) limits the formation of degradation products that inhibit the microbial fermentation.

6.5.1 Physical Pretreatments

Physical pretreatments such as chipping, ball and colloid milling, and grinding are mainly used to reduce size, increase surface area, and decrystallize (mechanically) biomass. It is not certain whether the benefits derived from the physical treatments are because of the reduction in crystallinity or the increase in surface area. Nonmechanical methods such as high temperature, freeze/thaw cycles, and irradiation have been attempted to change one or more structural features of cellulose and enhance hydrolysis. However, most of these methods are limited in their effectiveness and are often expansive.

6.5.2 Chemical Pretreatments

Chemical pretreatment methods employ the use of acid or alkali for fractionating biomass components, increasing surface area and pore volume, and swelling

Table 6.3. *Summary of important pretreatment methods (Sierra et al., 2008)*

Methods	Treatment time	Temperature (°C)	Pressure (bar)	Chemical used
Dilute acid (H_2SO_4)	5–30 min	140–190	4–13	0.5–1.5% H_2SO_4
Concentrated acid (H_3PO_4)	30–60 min	50	1	85% H_3PO_4
Peracetic acid ($C_2H_4O_3$)	1–1 80 h	25–75	1	2–10% $C_2H_4O_3$
Sodium hydroxide	24–96 h	25	1	1% NaOH, 0.1 g NaOH/g biomass
Lime – low lignin (12–18%)	24–96 h	100–120	1–2	0.10 g $Ca(OH)_2$/g biomass
Lime – medium lignin (18–24%)	~30 days	~55	1	0.10–0.15 g $Ca(OH)_2$/g biomass
Lime – high lignin (>24%)	2 h	~150	15	0.15–0.20 g $Ca(OH)_2$/g biomass
Ammonia recycle percolation	~15 min	180	~20	15% NH_3
Ammonia fiber explosion	~5 min	60–100	~20	1 g NH_3/g biomass
Oxidative alkali	6–25 h	20–60	1	NaOH (11.5 pH) + 1–15% H_2O_2 or O_3
Ionic liquid	5–12 h	50–150	–	Ionic liquids
Organosolv	60 min	180	35–70	1.25% H_2SO_4 60% ethanol
Steam explosion	0.3–50 min	190–250	12–40	Steam
Hydrothermal	1–15 min	160–220	6–35	Subcritical water
Supercritical CO_2	1 h	35–80	70–270	CO_2

cellulose. There are two types of cellulose swelling: (1) intercrystalline, which can be affected by water, and (2) intracrystalline, which can be affected by the use of chemical agents for breaking of hydrogen bonds. Some of the most commonly used pretreatment methods are discussed below.

6.5.2.1 Dilute Acids

Different acids, such as sulfuric acid, nitric acid, hydrochloric acid, phosphoric acid, and peracetic acid, have been used in the pretreatment process. Dilute acid pretreatment mainly removes the hemicelluloses fractions from the lignocellulosics. The removal of hemicelluloses increases the porosity in the biomass, and, thus, enzymatic accessibility to the cellulosic fractions is increased. Dilute acid is an efficient pretreatment method suitable for most lignocellulosic feedstock. Two approaches for dilute acid pretreatment are used. In one approach, high temperature (>160°C) in a continuous flow reactor with low solid loading is used. In another approach, a low-temperature batch process is used with high solid loading. In the batch operation, biomass is soaked with dilute acid for about 4 hours and then heated with directed steam injection. In the flow-through operation, hot dilute acid is pumped through a bed of biomass. Dilute sulfuric acid (0.5–1.5%) above 160°C was found to be most suitable for the industrial application, and this method fractionates the majority of hemicelluloses (75–90%).

The biggest advantage of dilute acid pretreatment is the fast rate of reaction, which facilitates continuous processing. Because five-carbon sugars degrade more rapidly than six-carbon sugars, one way to decrease sugar degradation is to have a two-stage process. The first stage is conducted under mild conditions to recover the five-carbon sugars. This is followed by the second stage, conducted under harsher conditions to recover the six-carbon sugars. The major disadvantage of this process is the removal of acids or neutralization, which yields a large amount of gypsum before the next step of enzymatic hydrolysis. Although close to theoretical sugar yields can be achieved, the process requires high capital cost coupled with corrosion problems, acid consumption, and recovery costs.

6.5.2.2 Peracetic Acid ($C_2H_4O_3$)

The use of peracetic acid (2–10%) at 25–75°C causes significant delignification of lignocellulosics due to the oxidizing action of peracetic acid. Reduction in crystallinity due to structural swelling and dissolution of crystalline cellulose are observed. In this pretreatment, the majority of hemicelluloses is retained in the solids. Drawbacks are the high cost and explosive nature of peracetic acids; as a result, this pretreatment has been limited to laboratory settings.

6.5.2.3 Concentrated Sulfuric Acid

Sulfuric acid is strong cellulose swelling and hydrolyzing agent. In the 62.5–70% concentration range, intracrystalline swelling of cellulose occurs, and above 75% concentration, dissolution and decomposition of cellulose occur. The dissolved cellulose is reprecipitated by the addition of methanol or ethanol. Upon pretreatment with 75% sulfuric acid, DP of cellulose drops from 2150 to 300 (Laxman and Lachke, 2008). Reprecipitated cellulose is easily hydrolyzed by dilute acid or enzyme with a high conversion to sugars. Sulfuric acid can be reused after the distillation of methanol or ethanol. However, large-scale testing of this process is still needed to determine the permissible recycling of sulfuric acid.

Concentrated acid hydrolysis requires relatively mild temperature and pressure with reaction time typically much longer than for dilute acid pretreatment. The concentrated sulfuric acid pretreatment can be followed by a dilution with water to hydrolyze the substrate into sugars. This process provides a complete and rapid conversion of cellulose to glucose and hemicelluloses to five-carbon sugars with little degradation.

Table 6.4 shows the yields of ethanol from cornstalks by concentrated sulfuric acid hydrolysis. The low-temperature and -pressure operation allow the use of relatively low-cost materials, such as fiberglass tanks and piping. The low-temperature and -pressure operation also minimize the degradation of sugars.

Unfortunately, concentrated acid is a relatively slow process, and cost-effective acid recovery systems have been difficult to develop. In one effort, acid and sugar are separated via ion exchange, and then acid is reconcentrated via multiple effect

Table 6.4. *Ethanol yield from corn stalks pretreated by concentrated sulfuric acid hydrolysis (Demirbas, 2005b)*

Basis: amount of corn stalk	1,000 kg
Cellulose content	430 kg
Cellulose conversion and recovery efficiency	76%
Ethanol stoichiometric yield	51%
Glucose fermentation efficiency	75%
Ethanol yield from glucose	125 kg
Hemicellulose content	290 kg
Hemicellulose conversion and recovery efficiency	90%
Ethanol stoichiometric yield	51%
Xylose fermentation efficiency	50%
Ethanol yield from xylose	66 kg
Total ethanol yield from 1,000 kg of cornstalks	191 kg (242 Lr = 64 gal)

evaporators. Without acid recovery, large quantities of lime must be used to neutralize the acid in the sugar solution. This results in the formation of large quantities of calcium sulfate, which requires disposal at an additional cost.

6.5.2.4 Concentrated Phosphoric Acid

Concentrated phosphoric acid (85%) has been applied earlier as a cellulose solvent. Phosphoric acid causes less degradation of the cellulose than other acids. Recently, a novel method for lignocelluloses fractionation was applied to hardwoods as well as softwoods. The major advantage of the method lies in the moderate processing condition (50°C and atmospheric pressure). Fractionation of lignocelluloses into highly reactive amorphous cellulose, hemicelluloses sugars, lignin, and acetic acid was achieved. After the pretreatment, enzymatic hydrolysis of Avicel and α-cellulose was completed within 3 hours while corn stover and switchgrass were hydrolyzed to the extent of 94% (Laxman and Lachke, 2008). Many of the limitations with concentrated phosphoric acids are the same as those discussed previously for sulfuric acid.

6.5.2.5 Ionic Liquids

Dissolution of cellulose in ionic liquid (e.g., 1-butyl-3-methylimidazolium chloride or BMImCl) and subsequent regeneration as amorphous cellulose with the use of an anti-solvent (e.g., water, ethanol, or methanol) has attracted some research interest. Hydrolysis of regenerated cellulose is significantly increased, and the initial rate of hydrolysis is approximately an order of magnitude higher than that of untreated cellulose. Nearly complete conversion of the carbohydrate fraction into water-soluble products is readily observed at 120°C in ionic liquids, which is much lower than the temperatures typically used for aqueous-phase hydrolysis (Sievers et al., 2009). Cost and recycling of ionic liquid are the major disadvantages of the process. In addition, the toxicity of various ionic liquids has not yet been fully established.

6.5.2.6 Alkali

Among the various chemical pretreatments, alkali pretreatment is the most widely used to enhance enzymatic hydrolysis of lignocellulose. Alkali pretreatment selectively removes the majority of lignin and part of the hemicellulose; the success of the method depends on the amount of lignin in the biomass. The main reagents used for alkali pretreatment are sodium hydroxide (NaOH), ammonia, calcium hydroxide, and oxidative alkali (NaOH + H_2O_2 or O_3). Dilute NaOH pretreatment causes disruption of lignin and carbohydrate linkages, swelling of cellulose, removal of lignin and hemicelluloses, increase in surface area, and decrease in the DP. The mechanism is considered to be saponification of intermolecular ester bonds cross-linking the hemicelluloses and lignin. The optimum levels range between 5 and 8 g NaOH/100 g substrate. The effectiveness of the pretreatment depends on the type of substrate. For example, the digestibility of softwood with high lignin content increased slightly compared with that of hardwoods (Millet, Baker, and Scatter, 1976). A cheaper alkali, calcium hydroxide, can also be used. For example, Karr and Holtzapple (2000) used calcium hydroxide as alkali for pretreating corn stover and poplar and applied a longer treatment time (a few days) at a low temperature. Similar to other alkaline pretreatments, calcium hydroxide causes a high delignification, resulting in an increase in the hydrolysis rate.

6.5.2.7 Ammonia

Pretreatment with aqueous ammonia was first patented in 1905. Ammonia causes a strong swelling action, which causes a crystal structure change from cellulose I to cellulose III_1. Ammonia shows high selectivity for lignin; as a result, the majority of hemicelluloses are retained in solids. In the batch process, termed soaking in aqueous ammonia, a high solid loading can be used at elevated temperature and pressure. In the continuous process, termed ammonia recycle percolation, an aqueous ammonia solution is fed to a column packed with biomass, and the solubilized lignin is continuously removed from the system so that lignin does not recondense onto the biomass. Relatively low cost and possibility of recycling due to its volatile nature are the major advantages of this process. In fact, the cost of ammonia, and especially of ammonia recovery, drives the cost of this pretreatment. However, biomass pretreatment economics are also strongly influenced by total sugar yields achieved and by the loss in yield and inhibition of downstream processes caused by sugar degradation products.

6.5.2.8 Organic Solvents

Pretreatment using organic solvents with mineral acids as catalysts have been effective in the delignification of biomass. Organic solvents break the lignin carbohydrate bonds selectively and fractionate the lignin and hemicelluloses. Solvents such as methanol, ethanol, acetone, ethylene glycol, triethylene glycol, and tetrahydrofurfuryl alcohol have been used. Mineral acids such as sulfuric acid and organic acids such as oxalic, acetylsalicylic, and salicylic acid can be used as catalysts. Organic solvents should be recycled for cost-effective use. Moreover, removal of organic solvents from the system after pretreatment is necessary because solvents may be inhibitory to the growth of organisms, enzymatic hydrolysis, and fermentation. From the organosolv process, lignin of high quality can be isolated and can therefore potentially add extra income to the biorefinery.

6.5.3 Hydrothermal Pretreatment

Hydrothermal pretreatment uses hot water at elevated pressure (more than saturation pressure) to ensure that water remains in liquid phase. Water acts as a solvent and as a reactant at the same time. Water as reactant is seen as a promising path to green chemistry by providing alternatives to corrosive acids and toxic solvents (Hashaikeh et al., 2007). The dielectric constant of water decreases with temperature, and water behaves like a more nonpolar solvent. The ionization constant of water increases with temperature below critical point. Such changes in properties of water with temperature lead to the cleavage of ether and ester bonds and favor the hydrolysis of hemicelluloses. Hydrothermal treatment of lignocellulosic biomass generates acid that arises from the thermally labile acetyl groups of hemicelluloses and catalyzes the hydrolysis of hemicelluloses and subsequent solubilization. Xylose recovery from biomass can be achieved as high as 88–98%. The structural alterations due to the removal of hemicelluloses increase the accessibility and enzymatic hydrolysis of cellulose. During the treatment, organic acids are librated from the biomass and pH of the reaction medium decreases. To avoid the formation of inhibitors, the pH should be kept between 4 and 7 during pretreatment; this in turn minimizes the formation of monosaccharides and therefore also the formations of degradation products that can further catalyze hydrolysis of the cellulosic material during pretreatment.

6.5.4 Physicochemical Pretreatments

6.5.4.1 Steam Hydrolysis and Explosion

High-pressure steaming, with or without rapid decompression (explosion), is considered one of the most successful options for fractionating wood into its three major components. Heating biomass in the presence of steam-saturated water (about 200°C and 15 bar, for typically 1–10 minutes) leads to organic acid generation due to the cleavage of hemicellulose acetyl groups. The evolved acid hydrolyzes some of the hemicelluloses and alters the lignin structure. Removal of hemicelluloses from the microfibrils is believed to expose the cellulose surface and increase enzyme accessibility to the cellulose microfibrils.

 If autohydrolysis is followed by rapid pressure release, the compressed water inside the biomass explosively vaporizes, which shatters the biomass with a popcorn-like effect and thus increases the surface area. This approach combines both chemical and physical pretreatments into one step, which is why steam explosion is a widely used method for pretreatment of biomass. The process is conducted typically between temperatures of 190 and 220°C and pressure 12 to 41 bar. The structural changes in biomass after the treatment, such as the removal and redistribution of hemicellulose and lignin, increase the pore volume. Rapid flashing to atmospheric pressure and turbulent flow fragment the material, thereby increasing the accessible surface area. The individual microfibrils do not seem to be affected significantly by the pretreatment. Depending on the severity of the pretreatment, some degradation of the cellulose to glucose also takes place. However, steam explosion does not always break down all of the lignin, requires small particle size, and also produces some inhibitory compounds for further enzymatic hydrolysis and fermentation steps. Sometimes the addition of sulfur dioxide (SO_2) or carbon dioxide (CO_2) during steam explosion treatment further improves the enzymatic hydrolysis of biomass by making the pretreatment environment more acidic (i.e., SO_2 forms sulfuric acid and CO_2 forms the carbonic acid in aqueous medium). The limitation of these methods is the lower yield of hemicellulose sugars. On the other hand, the low-energy input and negligible environmental effects are the major advantages of this process.

6.5.4.2 Ammonia Fiber Explosion (AFEX)

In the AFEX process, a batch of lignocelluloses is contacted with ammonia at high loadings and elevated temperature and pressure. After the reaction, the pressure is explosively released, similar to steam explosion, to disrupt the biomass structure. The physical disruption due to decompression and alkaline hydrolysis improves the enzymatic hydrolysis of biomass. Most of the ammonia (up to 99%) can be recovered for reuse. AFEX treatment does not solubilize much of the hemicelluloses and effectively results in higher sugar yield than acid-catalyzed steam explosion. AFEX was found very effective for pretreatment of low-lignin substrate (e.g., agriculture residue, herbaceous crops, and grasses) (Dale et al., 1996; Foster, Dale, and Doran-Peterson, 2001).

Unlike most other methods, no liquid fraction with dissolved products is generated by the AFEX pretreatment as ammonia is evaporated. Consequently, no lignin or other substances are removed from the material, but lignin–carbohydrate complexes are cleaved and deposition of lignin on the surface of the material is observed. Furthermore, AFEX results in depolymerization of the cellulose and partial hydrolysis of the hemicelluloses. Only little degradation of sugars occurs, and therefore low concentrations of inhibitors are formed. The AFEX method enables operating the process at high-solid concentrations. A continuous version of the AFEX process uses extruder and is termed fiber extrusion explosion.

6.5.4.3 Supercritical Carbon Dioxide

Supercritical CO_2 (critical point, 31°C and 74 bar) is a nontoxic and relatively cheaper pretreatment agent. In the presence of water, CO_2 creates carbonic acid effective for pretreatment. An explosive release of pressure fragments the biomass. The method has several advantages, such as low pretreatment temperatures, easy recovery of CO_2, and high solid loading. The most important effect is to decrease

the crystallinity. Although this process showed good results for Avicel, the low effectiveness for lignocellulosic biomass is the major drawback.

6.6 Cellulose Hydrolysis to Produce Sugars

Pretreatment increases the surface area and pore volume of biomass so that cellulose is more accessible for further hydrolysis. Depending on the pretreatment process used, lignin and/or hemicelluloses are removed and cellulose crystallinity is dis-
rupted. Cellulose is hydrolyzed using cellulase enzyme to produce sugars needed
for fermentation. Due to the high cost of enzyme, focus has been on improving cel-
lulases and decreasing the costs associated with the enzymatic hydrolysis of cellulose. Hydrolysis of cellulose requires the cooperation of three classes of cellulolytic
enzymes: (1) cellobiohydrolases, (2) endo-β-1,4-glucanases, and (3) β-glucosidases.
The CAZY (carbohydrate active enzymes; http://www.cazy.org) classification system collates glycosyl hydrolase enzymes into families according to sequence similar-
ity, which have been shown to reflect shared structural features (Demirbas, 2004a).
Some pretreatment methods leave the hemicelluloses in the material, so efficient
hydrolysis of these materials therefore also requires the use of another enzyme
hemicellulase. As hemicelluloses vary between different plant species, the optimal
enzyme mixture is most likely to be tailor made or adjusted to each different kind
of material.

In the process, the pH is adjusted and enzymes are added to initiate cellulose
hydrolysis to fermentable sugars. If hemicelluloses are also present, then additional
hemicellulase enzymes are added. Hydrolysis is typically performed at pH 5 and
50°C for 24–120 hours, which is followed by addition of a fermentation organism
to begin production of ethanol. In many cases, fermentation is started long before
hydrolysis has completed, since both the extent and speed of ethanol production can
often be increased by combining the hydrolysis and fermentation steps (Jørgensen
et al., 2007).

The challenges with enzymatic hydrolysis of lignocellulose include enzyme cost,
product inhibition, loss of enzyme due to binding with lignin, and denaturation
or degradation. To reduce the enzyme cost, enzyme loading should be minimized.
Lower enzyme loading, however, increases the time needed to complete hydrolysis.
Also, the use of high substrate concentrations increases the problem of product inhibition, which results in lower performance of the enzymes. Lignin, if present, shields
the cellulose chains and adsorbs the enzymes, decreasing the efficiency of hydrolysis. Furthermore, the activity of some enzymes might be lost due to degradation or
denaturation. Other factors are closely linked to the substrate composition and thus
the pretreatment method employed (Galbe et al., 2007). In-house production of the
enzyme using a fungus near the hydrolysis site is a viable option. Specifically, one
can use part of the biomass to run the enzyme production unit, and crude enzyme
itself can be used in the hydrolysis unit.

The amount and types of enzymes required strongly depend on the biomass
being hydrolyzed and the type and severity of pretreatment used. The selection

of biomass feedstock is based on local availability and cost. With variation in the biomass, different thermochemical pretreatments should be used to balance accessibility to enzymatic attack without destruction of valuable sugars. Variations in severity (combined effect chemical concentration, temperature, acidity, pressure, and duration of treatment) of the pretreatment should maximize both sugar and fermentation compatibility. For example, a low-severity pretreatment will solubilize less of the hemicellulose fraction, increasing the amount of hemicellulase enzymes required, but may also reduce the production of byproducts toxic to the fermentation, increasing the overall ethanol yield.

The composition of each biomass varies depending on the plant species, age at harvesting, and local soil and climate. The key substrate characteristics that impact the rate of hydrolysis include accessibility, degree of cellulose crystallinity, and the type and distribution of lignin. The presence of lignin is hypothesized to decrease the quantity of the enzyme available for cellulose due to nonspecific adsorption of the enzyme to lignin and steric hindrance when lignin encapsulates the cellulose. Each of these factors is known to effect enzyme action, and no single parameter correlates absolutely with the enzymatic digestibility. Hence, the variation in composition of a given biomass requires some tailoring in the hydrolysis process (Merino and Cherry, 2007).

6.7 Fermentation of Sugars to Ethanol

Hydrolysis of cellulose and hemicelluloses produces six-carbon and five-carbon sugars, respectively. To produce ethanol from six-carbon sugars, one can use a technology similar to the one used for corn and sugarcane feedstocks (described in Chapter 5).

The process in which hydrolysis is followed by fermentation is called the separate hydrolysis and fermentation (SHF) process. Here, the stream from the hydrolysis process passes on to a fermenter to which yeast is added to convert the glucose into ethanol. The glucose yield in SHF is typically low due to end-product inhibition of the hydrolysis step by glucose and cellobiose. Another processing option is simultaneous saccharification and fermentation (SSF), which results in a higher ethanol yield attributed to the removal of glucose and cellobiose by the fermentation avoiding the hydrolysis inhibition. Additional benefits include the following: (1) glucose separation from the lignin fraction following a separate enzymatic hydrolysis step is not needed, thereby avoiding a potential loss of sugar; (2) the combination of two operations decreases the number of vessels needed and thereby capital costs (a savings of at least 20%, which is close to the biomass feedstock cost); (3) co-consumption of pentose and hexose sugars; and (4) ease of detoxification.

However, there are some disadvantages of the SSF process as compared with the SHF process. The optimum temperature for enzymatic hydrolysis is typically higher than that for fermentation using yeast. Hence, one needs to use some compromise conditions for temperature and pH suitable for both hydrolysis (saccharification) and fermentation. The recycling of enzymes and fermentation organism is difficult, which is further exacerbated by the problem of separating yeast from the lignin after fermentation. Hence, one is forced to use lower yeast loadings. Recycling of enzymes is equally difficult in both SSF and SHF processes because the enzymes bind to the substrate; although, a partial desorption can be obtained upon addition of surfactants. The enzymes are either produced within the process, thereby representing a loss of substrate, or are externally supplied and thereby add to the input costs.

Because five-carbon sugars (xylose) comprise a high percentage of the available sugars, the ability to recover and ferment them into ethanol is important for the efficiency and economics of the process.

6.7.1 Xylose Fermentation

D-Xylose ($C_5H_{10}O_5$), also known as wood sugar, is an aldopentose – a monosaccharide containing five carbon atoms with an aldehyde functional group. The metabolic steps involved in the fermentation of five-carbon sugars have not been extensively studied compared with six-carbon sugars. The first biochemical step in D-xylose fermentation is the isomerization to a keto isomer, D-xylulose. The conversion in bacteria is performed by D-xylose isomerase enzyme, and that in yeast by a two-step enzymatic reaction involving reduction and oxidation (i.e., NADPH-linked D-xylose reductase converts D-xylose to Xylitol, which is further converted to D-xylulose by NAD-linked xylitol dehydrogenase). Next, D-xylulose is phosphorylated by D-xylulokinase for assimilation via the pentose phosphate pathway. Conversion by D-xylulokinase is the chief driving reaction in the pathway, which proceeds by way of epimerase, isomerase, transketolase, and transaldolase. Requirements for the induction of these enzymes and their proportion, as well as the availability of the necessary cofactors, are thought to be the major factors influencing pentose fermentation.

Bacteria have drawn special attention from researchers because of their speed of fermentation. In general, bacteria can ferment in minutes compared with hours for yeast. Many bacteria are able to ferment xylose, especially members of the genera *Clostridium* and *Bacillus* (*Bacillus macerans*) in which ethanol is a major end-product. However, the main problem in the use of bacteria is their low ethanol tolerance and the formation of other byproducts. The rate and yield of ethanol production from D-xylose depend on several process variables, including D-xylose concentration, bacteria used, nitrogen source, aeration, and agitation. For example, *Candida shehatae* is an efficient fermenter of D-xylose but is not able to use nitrate as the sole nitrogen source. On the other hand, in the case of *Pachysolen tannophilus*, oxygen is required for cell growth but not for ethanol production. Each gram of D-xylose, using *P. tannophilus*, yields 0.34 g ethanol as compared with the theoretical yield of 0.51 g (Lachke, 2002).

The ethanol production rate and final yield from D-xylose fermentation are still much lower than those from glucose fermentation on a commercial scale. The reasons for low rate and yield include formation of other products (xylitol, acetate, etc.) and the use of nonoptimized fermentation, as parameters (e.g., aeration, pH, C:N ratio, various additives, and respiratory inhibitors) influencing the fermentation process have not yet been fully evaluated. Upon optimization, it is expected that the production of 5–6% (w/v) ethanol will be attained. Hence, the fermentation of D-xylose is only economically feasible when it is a side-process of lignocellulose processing. Adaptation of yeasts for fermentation using waste streams or hydrolysates is another essential feature necessary for process development (Lachke, 2002; Agbogbo and Coward-Kelly, 2008). Recent results with *P. stipitis* yeast show encouraging results, that it can ferment both glucose and xylose (Agbogbo et al., 2008; Huang et al., 2009).

6.8 Ethanol Separation and Purification

The ethanol separation and purification process is similar to that for corn or sugarcane feedstock, as discussed in Chapter 5.

6.9 Summary

Starch or sugar-based ethanol production has been blamed for the rise in the food prices. To satisfy current and future demands, ethanol production from lignocellulosic biomass fermentation is a viable option that does not compete with the food supply. The process includes the following key steps: pretreatment, hydrolysis of cellulose, hydrolysis of hemicelluloses, fermentation of five-carbon and six-carbon sugars, separation of lignin residue, and recovery and concentration of ethanol. Currently, the ethanol produced from lignocellulosics is more expensive than that from starches. Hence, production costs need to be reduced. Requirements for this cost reduction include effective pretreatment to reduce cellulase use, hydrolysis of hemicellulose and cellulose to sugars, use of both six-carbon and five-carbon sugars, and process integration for reducing capital and energy costs.

7 Biodiesel from Vegetable Oils

7.1 What Is Biodiesel?

Biodiesel refers to a renewable fuel for diesel engines that is derived from animal fats or vegetable oils (e.g., rapeseed oil, canola oil, soybean oil, sunflower oil, palm oil, used cooking oil, beef tallow, sheep tallow, and poultry oil). Biodiesel is a clear amber–yellow liquid with a viscosity similar to petroleum diesel (petrodiesel, diesel). With the flash point of 150°C, biodiesel is nonflammable and nonexplosive, in contrast to petrodiesel, which has a flash point of 64°C. This property makes biodiesel-fueled vehicles much safer in accidents than those powered by diesel or gasoline. Unlike petrodiesel, biodiesel is biodegradable and nontoxic and significantly reduces toxic and other emissions when burned as a fuel. Technically, biodiesel is a diesel engine fuel comprised of monoalkyl esters of long-chain fatty acids derived from animal fats or vegetable oils, designated B100, and meeting the requirements of the ASTM D-6751 standard. Some of its technical properties are listed in Table 7.1. Chemically, biodiesel is referred to as a monoalkyl ester, especially (m)ethylester, of long-chain fatty acids derived from natural lipids via the transesterification process. Biodiesel is typically produced by reacting a vegetable oil or animal fat with methanol or ethanol in the presence of a catalyst to yield methyl or ethyl esters (biodiesel) and glycerin (Demirbas, 2002a). Generally, methanol is preferred for transesterification, because it is less expensive than ethanol.

Biodiesel produces slightly lower power and torque, which results in a higher consumption than No. 2 diesel fuel for driving. However, biodiesel is better than diesel in terms of sulfur content, flash point, aromatic content, and biodegradability. Precautions should be taken in very cold climates, where biodiesel may gel earlier than diesel upon cooling.

The cost of biodiesels varies depending on the feedstock, geographic area, methanol prices, and seasonal variability in crop production.

7.2 History

The emergence of transesterification dates back to as early as 1846 when Rochieder described glycerol preparation through ethanolysis of castor oil. Seven years later, in

Table 7.1. *General properties of biodiesels*

Chemical name	Fatty acid (m)ethyl ester
Chemical formula range	C_{14}–C_{24} methyl esters or C_{15-25} $H_{28-48}O_2$
Kinematic viscosity range	3.3–5.2 mm^2/s, at 40$_\circ$C
Density range	860–894 kg/m^3, at 15°C
Boiling point range	200°C
Flash point range	155–180°C
Distillation range	195–325°C
Vapor pressure	<5 mm Hg, at 22°C
Solubility in water	Insoluble in water; however, biodiesel can absorb up to 1500 ppm water
Physical appearance	Light to dark yellow, clear liquid
Odor	Light musty/soapy odor
Biodegradability	More biodegradable than petrodiesel
Reactivity	Stable, but reacts with strong oxidizers

1853, scientists E. Duffy and J. Patrick conducted transesterification of a vegetable oil, long before the first diesel engine became functional. German inventor Rudolph Diesel designed the diesel engine in 1893 with a revolutionary design in which air could be compressed by a piston to a very high pressure, thereby causing a high temperature suitable for combustion. The original diesel engine used peanut oil. Diesel believed that the use of a biomass fuel was the real future of his engine. In a 1912 speech, he said, "the use of vegetable oils for engine fuels may seem insignificant today, but such oils may become, in the course of time, as important as petroleum and the coal-tar products of the present time."

The use of vegetable oils as alternative renewable fuel, competing with petroleum, was proposed in the early 1980s, but commercial production did not begin until the late 1990s. Since the 1980s, biodiesel plants have opened in many European countries, and some cities have run buses on biodiesel or blends of petro- and biodiesels. Recent environmental and domestic economic concerns have prompted resurgence in the use of biodiesel throughout the world. In 1991, The European Community proposed a 90% tax deduction for the use of biofuels, including biodiesel. Biodiesel plants are now being built by several companies in Europe; each of these plants will produce up to 1.5 million gallons of fuel per year. The European Union accounted for nearly 89% of all biodiesel production worldwide in 2005.

7.3 Vegetable Oil Resources

Vegetable oils have different chemical structures than diesel. Vegetable oils contain fatty acids that are linked to a glycerin molecule with ester linkages, called a triglyceride (Figure 7.1). The fatty acids are in their carbon chain length and in numbers of double bonds.

As can be seen in Table 7.2, palmitic (16:0) and stearic (18:0) acids are two of the most common saturated fatty acids found in every vegetable oil. Similarly, oleic

Table 7.2. *Fatty acid composition of vegetable oils*

Fatty acid $(xx{:}y)^a \rightarrow$	16:0	16:1	18:0	18:1	18:2	18:3	Others
Palm	42.6	0.3	4.4	40.5	10.1	0.2	1.1
Soybeans	11.9	0.3	4.1	23.2	54.2	6.3	0
Rapeseed	3.8	0	2.0	62.2	22.0	9.0	0
Sunflower seed	6.4	0.1	2.9	17.7	72.9	0	0
Peanut[b]	11.4	0	2.4	48.3	32.0	0.9	4.0
Cottonseed	28.7	0	0.9	13.0	57.4	0	0
Coconut[c]	7.8	0.1	3.0	4.4	0.8	0	65.7
Olive	5.0	0.3	1.6	74.7	17.6	0	0.8

[a] xx is the number of carbon atoms and y is the number of carbon–carbon double bonds in the fatty acid; an oil with higher y is more unsaturated and considered healthier for human consumption.
[b] Peanut oil contains about 2.7% of 22:0 and 1.3% of 24:0 fatty acids.
[c] Coconut oil contains about 8.9% of 8:0, 6.2% of 10:0, 48.8% of 12:0, and 19.9% of 14:0 fatty acids.

(18:1) and linoleic (18:2) acids are the most common unsaturated fatty acids. Many of the oils also contain some linolenic acid (18:3).

There is little variation between gross heat content among the vegetable oils. Table 7.3 compares fuel properties of vegetable oils with No. 2 diesel fuel. The heat content of vegetable oil is approximately 88% of No. 2 diesel fuel.

Densities of vegetable oils are between 912 and 921 kg/m^3, whereas that of No. 2 diesel fuel is 815 kg/m^3. The kinematic viscosities of vegetable oils vary between 39.2 and 65.4 mm^2/s at 27°C, which are 9–15 times higher than that for No. 2 diesel fuel. Hence, vegetable oils are extremely viscous.

World vegetable oil production of about 0.13 billion tons/year is very small when compared with the world petroleum consumption at 4.25 billion tons/year. Hence, vegetable oils can only substitute a very small fraction of petroleum-based engine fuels in the near future (Demirbas, 2003; Bala, 2005). Furthermore, some environmental groups object to the significant increase in farming, which will result in over-fertilization, pesticide use, and land-use conversion. Many advocates suggest

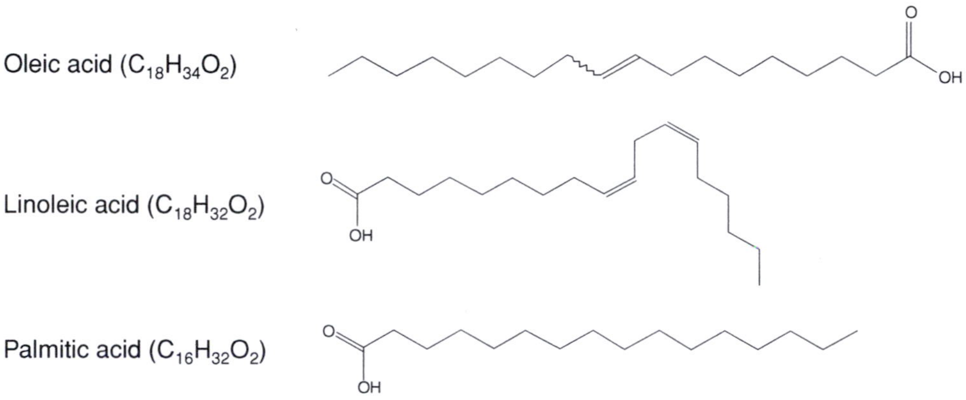

Oleic acid (C$_{18}$H$_{34}$O$_2$)

Linoleic acid (C$_{18}$H$_{32}$O$_2$)

Palmitic acid (C$_{16}$H$_{32}$O$_2$)

Figure 7.1. Major fatty acid components in the vegetable oils.

Table 7.3. *Comparisons of fuel properties of vegetable oils with No. 2 diesel fuel*

Fuel type	Heating value (MJ/kg)	Density (kg/m^3)	Viscosity at 27 °C (mm^2/s)	Cetane number[a] (CN)
No. 2 diesel fuel	43.4	815	4.3	47.0
Sunflower oil	39.5	918	58.5	37.1
Cottonseed oil	39.6	912	50.1	48.1
Soybean oil	39.6	914	65.4	38.0
Corn oil	37.8	915	46.3	37.6
Poppy oil	38.9	921	56.1	–
Rapeseed oil	37.6	914	39.2	37.6

[a] Cetane number (CN) is a measure of ignition quality of the fuel in a diesel engine.

that waste vegetable oil is the best source of oil to produce biodiesel. However, the available supply is significantly less than the amount of petroleum-based fuel that is burned for transportation and home heating in the world. Animal fats are similarly limited in supply, and it would not be efficient to raise animals simply for their fat. However, producing biodiesel with animal fat that would have otherwise been discarded could replace a small percentage of petroleum diesel usage.

A variety of oils and fats can be used to produce biodiesel, including (1) virgin vegetable oils, such as palm, soybean, and algae oils; (2) waste vegetable oil; (3) animal fats, such as tallow, lard, and yellow grease; and (4) nonedible oils, such as jatropha, neem oil, castor oil, tall oil, and so on. There are more than 350 identified oil-bearing crops, among which only soybean, palm, sunflower, safflower, cottonseed, rapeseed, and peanut oils are considered as potential feedstocks for biodiesel. Table 7.4 shows the world vegetable consumption between 2000 and 2008, with palm oil and soybean oil contributing to the majority of consumption.

Prominent feedstock for biodiesel production in Malaysia and Indonesia is palm oil, in Europe is rapeseed oil, and an emerging source in India is jatropha oil. Soybeans are commonly used in the United States for food products, which has led to soybean biodiesel becoming the primary source for biodiesel in that country.

Table 7.4. *World vegetable oil consumption (in million metric tons/year)*

Oil	2000	2002	2004	2007	2008
Palm	23.3	25.8	30.5	40.8	40.2
Soybeans	26.0	30.5	31.6	38.4	37.5
Rapeseed	13.1	11.6	15.5	18.0	18.4
Sunflower seed	8.6	8.3	8.6	10.1	9.0
Peanut	4.2	4.4	5.0	5.0	5.0
Cottonseed	3.6	3.5	4.6	4.9	4.8
Palm kernel	3.3	3.4	3.6	4.8	4.8
Coconut	2.7	3.2	3.4	3.3	3.4
Olive	2.5	2.5	2.7	3.0	2.9
Total	87.2	93.2	105.6	128.2	125.8

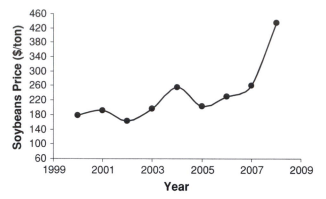

Figure 7.2. U.S. soybean prices.

Biodiesel use has resulted in a significant rise in soybean prices (Figure 7.2). Such increases in oil seeds have given rise to the food-versus-fuel debate. Hence, the focus is shifting toward using nonedible vegetable oils that are produced from marginal lands.

7.3.1 Nonedible Oil Resources

The main nonedible oil resources for biodiesel include jatropha, ratanjot, or seemai-kattamankku (*Jatropha curcas*), karanja or honge (*Pongamia pinnata*), nag champa (*Calophyllum inophyllum*), rubber seed tree (*Hevea brasiliensis*), neem (*Azadirachta indica*), mahua (*Madhuca indica* and *Madhuca longifolia*), silk cotton tree (*Ceiba pentandra*), jojoba (*Simmondsia chinensis*), babassu tree, milk bush (*Euphorbia tirucalli*), algae, and so on. Jatropha and karanja, found in India, have a high oil content (25–30% of seed). In fact, jatropha oil was used as a diesel fuel substitute during World War II (Shah, Sharma, and Gupta, 2004). Plant for rubber seed oil (20–30% of seed) is found mainly in Indonesia, Malaysia, Liberia, India, Sri Lanka, Sarawak, and Thailand. Oil palms, *Elaeis guineensis* and *Elaeis oleifera*, are found in Africa and Central/South America, respectively. Evergreen neem tree is found in tropical and semitropical regions of the Indian subcontinent. Jojoba oil is produced from the seed of the jojoba plant, a shrub native to Southern Arizona, Southern Calfornia, and Northwestern Mexico. The silk cotton tree has great economic importance for both domestic and industrial uses in Nigeria, and its oil is a good source for biodiesel production. Another nonedible oil is obtained from the babassu tree, which is widely grown in Brazil (Srivastava and Prasad, 2000).

Algae can grow in practically any place where there is enough sunshine, including areas with saline water. The most significant difference from the oil crop is the yield of algal oil. According to some estimates, the per-acre yield of oil from algae is more than 200 times the yield from the best-performing vegetable oil plant (Sheehan et al., 1998). About 19 tons of oil/acre/year can be produced from diatom algae, a type of fast-growing microalgae that can complete an entire growing cycle every few days. However, the production of algae to harvest oil for biodiesel has not been undertaken on a commercial scale, and the commercial-scale oil production may be less. Some algae produce up to 50% oil by weight with fatty acid composition of

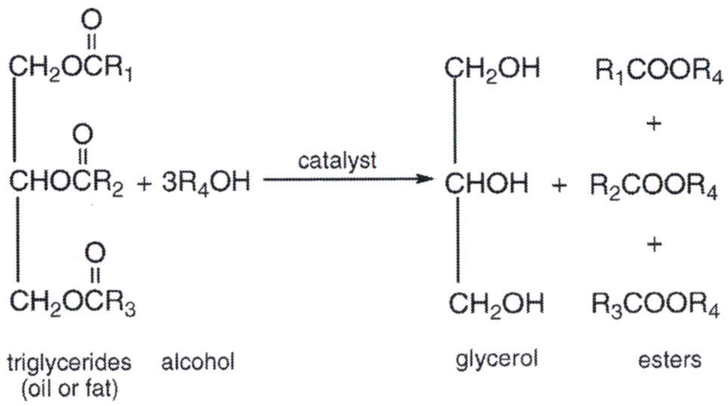

Figure 7.3. Transesterification of a tri-glyceride with alcohol (Pinzi et al., 2009; reproduced with permission from ACS).

36% oleic (18:1), 15% palmitic (16:0), 11% stearic (18:0), 8.4% iso-heptadecanoic acid (17:0), and 7.4% linoleic (18:2) acid. The high proportion of saturated and monounsaturated fatty acids in this alga is considered optimal from a fuel, in that fuel polymerization during combustion would be substantially less than what would occur with polyunsaturated fatty acid-derived biodiesel (Sheehan et al, 1998).

7.4 Transesterification

Fatty acids in oils are present in the form of triglycerides. For example, palmitic acid, oleic acid, and α-linolenic acid can be joined by a glycerol unit to form the following triglyceride:

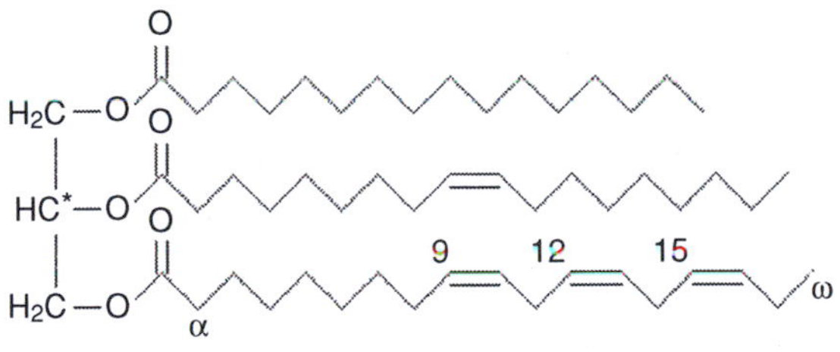

In transesterification (or alcoholysis), triglyceride is reacted with an alcohol (e.g., methanol or ethanol) to form esters and glycerol, as shown in Figure 7.3. A catalyst (e.g., KOH, NaOH) is usually used to enhance rate and yield of the reaction. Because the reaction is reversible, excess alcohol is used to shift the equilibrium to the product side.

7.4.1 Catalytic Methods

Transesterification can be catalyzed by alkalis, acids, or enzymes (Zhang et al., 2003; Noureddini, Gao, and Philkana, 2005). A typical procedure for the alkali-catalyzed method is as follows. The catalyst (KOH or NaOH) is dissolved into methanol by vigorous stirring in a small vessel. Then, this mixture is pumped into a reactor containing oil. The reactor is heated (71°C) and vigorously stirred for about 2 hours to complete the transesterification. Upon successful completion of the reaction, settling of the phases is allowed in which crude glycerin (heaver liquid) collects at the bottom and biodiesel (lighter liquid, ester) collects at the top. The phase separation

Table 7.5. *Comparisons of various methanolic transesterification methods*

Method	Reaction temperature (°C)	Reaction time (min)
Acid or alkali catalytic process	30–70	60–360
Boron trifluoride–methanol	87–117	20–50
Sodium methoxide–catalyzed	20–25	4–6
Noncatalytic supercritical methanol	250–300	6–12
Catalytic supercritical methanol	250–300	0.5–1.5

can start in 10 minutes, but complete settling can take 2–20 hours. After settling is complete, ester is carefully washed. Water is added at 5.5% by volume of the ester and then stirred for 5 minutes, and the glycerin is allowed to settle again. A water wash solution at 28% by volume of ester and 1 g of tannic acid per liter of water is added to the ester and gently agitated. Air is carefully introduced into the bottom aqueous layer while simultaneously stirring very gently. The process is continued until the ester layer becomes clear. After settling, the aqueous solution is drained, and water is added at 28% by volume of ester for the final washing (Ma and Hanna, 1999; Demirbas, 2002a). Now, the aqueous phase is again drained, resulting in the final biodiesel product.

For sodium methoxide-catalyzed transesterification, vegetable oil is transesterified in a solvent (e.g., toluene) with methanol containing fresh sodium. The reaction is typically carried out at 25°C for 10 minutes.

In the case of acid catalysis, sulfuric acid, hydrochloric acid, and sulfonic acid are usually preferred as acid catalysts. The catalyst is dissolved into methanol by vigorous stirring in a small vessel. Then, the mixture is pumped into a biodiesel reactor containing vegetable oil. The reaction is typically carried out at 30–35°C for 1–6 hours. Various transesterification methods are compared in Table 7.5.

A major challenge with the above transesterification is that free fatty acids in the presence of water cause soap formation, which reduces catalyst effectiveness and results in a low conversion. In addition, the separation and cleaning of the product and catalyst result in high material and energy costs.

7.4.2 Noncatalytic Supercritical Alcohol Method

In order to overcome the problems of conventional transesterification, Kusdiana and Saka (2001) and Demirbas (2002a, 2003) have proposed esterification of vegetable oil with supercritical methanol (SCM). This novel, noncatalytic, one-phase (due to a low dielectric of SCM) process has solved some of the challenges arising from the two-phase nature of normal methanol/oil mixtures. In the SCM process, the reaction completes in a very short time and the purification of the product is much simpler. However, the reaction requires temperatures of 250–400°C and pressures of 350–600 bar (Kusdiana and Saka, 2001; Demirbas, 2003; Saka and Kusdiana 2001).

Other alcohols can also be used, and their critical properties are shown in Table 7.6. However, the rate of conversion in higher alcohols is slower than that

Table 7.6. *Critical temperatures and critical pressures of various alcohols*

Alcohol	Critical temperature (°C)	Critical pressure (bar)
Methanol	239	81
Ethanol	243	64
1-Propanol	264	51
1-Butanol	287	49

in methanol (Figure 7.4). In addition, typically, the higher alcohols are more expensive than methanol.

In the transesterification process, vegetable oil should have an acid value of less than 1, and all the materials should be substantially anhydrous. The higher the acid value, the more NaOH or KOH is required to neutralize the free fatty acids. The absence of water can prevent soap formation and frothing (Demirbas, 2003). Table 7.7 compares catalytic methanol and supercritical methanol methods.

7.4.3 Recovery of Glycerol

About 10 tons of glycerol is produced for every 100 tons of biodiesel. Hence, a worldwide production of 13 million tons/year of biodiesel has given rise to 1.3 million tons/year of glycerol. Recovery and use of this byproduct can add value to the biodiesel process. However, a recent surge of glycerol supply in the market has significantly depressed its prices. In a steady economic operation, valuable products can be made from glycerol; hence, its efficient recovery is important. The separation

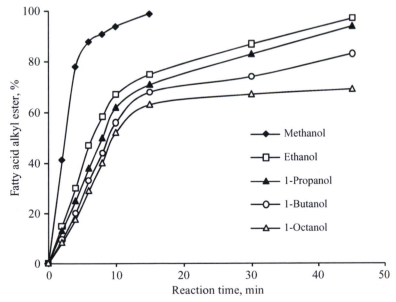

Figure 7.4. Production of fatty acid alkyl esters from triglycerides in supercritical alcohol at 300°C (adapted from Demirbas, 2008a).

Table 7.7. *Comparisons of catalytic methanol (MeOH) and supercritical methanol (SCM) methods for biodiesel production (Demirbas, 2008a)*

	Catalytic MeOH process	SCM method
Methylating agent	Methanol	Methanol
Catalyst	Alkali (NaOH or KOH)	None
Reaction temperature (K)	303–338	523–573
Reaction pressure (MPa)	0.1	10–25
Reaction time (min)	60–360	7–15
Methyl ester yield (wt%)	96	98
Removal for purification	Methanol, catalyst, glycerol, soaps	Methanol, glycerol
Free fatty acids	Saponified products	Methyl esters, water
Smell from exhaust	Soap smelling	Sweet smelling

of biodiesel and glycerin can be achieved using an inexpensive settling tank. If faster separation is needed, then a centrifuge can be used. The denser phase preferentially separates to the outer surface of the centrifuge. The choice of appropriate centrifuge type and size is dependent on the degree of separation needed in a specific system.

Glycerol has been used in a variety of ways, including as a humectant, plasticizer, emollient, thickener, dispersing medium, lubricant, sweetener, bodying agent, and antifreeze. Its most popularly use has been as an additive in cosmetics, toiletries, personal care, drugs, and food industries. In addition, glycerol has been employed as a raw material in various chemical syntheses, including production of polyethers, propylene glycols (specialy 1,3-propanediol for polyester manufacturing), dihydroacetone, and glyceraldehyde (Wolfson et al., 2009). Unfortunately, in all of the applications mentioned above, highly refined and pure glycerol is required. Biodiesel manufacturing produces somewhat "dirty" glycerol, which will need to go through purification steps, including bleaching, deodorizing, and ion exchange, to remove trace impurities.

Usually, these steps are very expensive. Hence, economical uses for low-grade glycerol need to be explored to further defray the cost of biodiesel production.

7.4.4 Reaction Mechanism

Transesterification consists of three consecutive, reversible reactions (Equations 7.1–7.3). In stepwise fashion, the triglyceride is converted to diglyceride, monoglyceride, and, finally, glycerol (Equation 7.3), producing a molecule of alkyl esters in each step. The formation of alkyl ester from monoglyceride is believed to be the slowest step due to the high stability of the monoglyceride intermediate (Ma and Hanna, 1999).

$$\text{Triglyceride} + \text{ROH} \leftrightarrow \text{Diglyceride} + \text{RCOOR}_1 \tag{7.1}$$

$$\text{Diglyceride} + \text{ROH} \leftrightarrow \text{Monoglyceride} + \text{RCOOR}_2 \tag{7.2}$$

$$\text{Monoglyceride} + \text{ROH} \leftrightarrow \text{Glycerol} + \text{RCOOR}_3 \tag{7.3}$$

The course of transesterification is affected by several factors, including the type of catalyst (alkaline, acid, or enzyme), molar ratio of alcohol to vegetable oil, temperature, purity of the reactants (mainly water content), and free fatty acid content. In conventional transesterification, free fatty acid and water always produce negative effects, as the presence of these compounds cause soap formation, consumes catalyst, and reduces catalyst effectiveness, leading to a low conversion (Kusdiana and Saka, 2004a-b). The alcohol-to-vegetable oil molar ratio plays a major role in transesterification. For example, an excess of alcohol favors the formation of ester products. However, a further increase in the amount of alcohol makes the recovery of the glycerol difficult. Hence, an optimum alcohol-to-oil ratio has to be established for a given process.

Base-catalyzed transesterification is faster than acid-catalyzed. The base catalyst first reacts with the alcohol, producing an alkoxide and the protonated catalyst. The alkoxide attacks the carbonyl group of the triglyceride, generating a tetrahedral intermediate, from which the alkyl ester and the corresponding anion of the diglyceride are formed. The anion deprotonates the catalyst, thus regenerating it for another cycle. Diglycerides and monoglycerides are converted by the same mechanism to alkyl esters and glycerol.

Alkaline metal alkoxides (e.g., CH_3ONa and CH_3CH_2ONa) are the most active catalysts because they give very high yields (>98%) in short reaction times (~30 minutes), even when applied at low molar concentrations (0.5 mol%). However, these alkoxides require the absence of water, which makes them inappropriate for typical industrial processes because commercial vegetable oils do contain some amount of moisture. On the other hand, alkaline metal hydroxides (e.g., KOH and NaOH) are cheaper than metal alkoxides, but are less active.

Water present in alcohol or oil can hydrolyse some of the produced ester, with consequent soap formation. This undesirable saponification reduces the biodiesel yields and causes difficulty in glycerol recovery due to the formation of emulsions. Some remediations have been developed. For example, the use of 2–3 mol% potassium carbonate can increase the yield and reduce soap formation.

7.5 Current Technologies

Most of the biodiesel produced today is made via the base-catalyzed reaction for several reasons.

- It requires low temperature and pressure.
- It provides a high yield (98% conversion) with minimal side reactions and reaction time.
- It directly converts oil to biodiesel with no intermediate compounds.
- The reactors can be made from simple materials (e.g., stainless steel).

A schematic of the process is shown in Figure 7.5 with the following key steps: (1) mixing of alcohol and catalyst, (2) transesterification reaction, (3) separation,

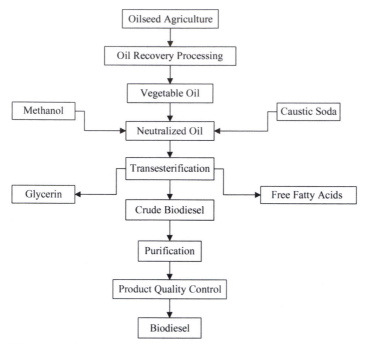

Figure 7.5. Simplified flow diagram of base-catalyzed biodiesel processing.

(4) biodiesel washing, (5) unreacted alcohol removal, (6) glycerin neutralization, and (7) the final product refinement.

7.5.1 Raw Materials and Feedstock Preparation

Choosing the oils or fats to be used in producing biodiesel is an important aspect of the economic decision, as the cost of oils or fats can account for up to 70–80% of the total cost. Preferred vegetable oil sources are soybean, canola, palm, and rape. Preferred animal fat sources are beef tallow, lard, poultry fat, and fish oils. Yellow greases, which are mixtures of oils and fats, can also be used. Waste cooking oil presents an attractive feedstock as its use can bring a good value to an otherwise waste material that would need to be properly discharged (Zhang et al., 2003). Crude vegetable oils contain some phospholipids and free fatty acids, which are removed by degumming and refining processes, respectively.

The second most important material to choose is the alcohol. The most common choice is methanol, although other alcohols, such as ethanol, isopropanol, and butyl alcohol, can also be used. An important factor in alcohol selection is the water content, as water interferes with transesterification reactions and can result in poor yield and high levels of soap, free fatty acids, and triglycerides in the final biodiesel fuel. A water content of merely 0.5 wt% can stop the reaction; hence, water content should be much less than 0.5 wt%.

The theoretical (stoichiometric) requirement for alcohol is three moles for each mole of triglyceride. The reaction rate and yield can be enhanced by using a higher

Table 7.8. *Typical proportions for the chemicals used to make biodiesel*

Reactants	Amount (kg)
Fat or oil	100
Primary alcohol (methanol)	10
Catalyst (sodium hydroxide)	0.30
Neutralizer (sulfuric acid)	0.36

than 3:1 mole ratio. The commonly used ratio in the industry is 6:1, which results in the following material balance:

$$\text{Triolein} + 6\,\text{Methanol} \rightarrow \text{Methyl oleate} + \text{Glycerol} + 3\,\text{Methanol} \qquad (7.6)$$
$$885.46\text{g} \qquad 192.24\text{g} \qquad\qquad 889.50\text{g} \qquad 92.10\text{g} \qquad 96.12\text{g}$$

The methanol that remains after the reaction is separated and used in the next cycle. The main materials needed are listed in Table 7.8.

7.5.1.1 Choice of Alcohol
The choice of alcohol depends on issues such as cost, the amount needed for the reaction, the ease of recovering and recycling, fuel tax credits, and global warming potential. Use of some alcohols requires modifications in the operating temperatures, mixing times, and/or mixing speeds. Theoretically, it takes three moles of alcohol to completely react with each mole of triglyceride. But the commercial sale of alcohol is based on volume or mass rather than moles. Hence, a proper economic evaluation is needed when selecting an alcohol. For example, to process 1 ton of triolein, 109 kg of alcohol is required when using methanol, but 156 kg alcohol is required when using ethanol. Two common choices of alcohols are methanol and ethanol. Sources of methanol include natural gas, petroleum gas, coal pyrolysis, wood pyrolysis, and wood gasification. The sources for ethanol include fermentation of sugars and starches, cellulosic biomass, synthesis from petroleum, synthesis from coal, and so on (Demirbas, 2008a). Most commercial methanol comes from petroleum sources, whereas ethanol is from sugars and starches. Typically, methanol is the preferred alcohol as it is much cheaper than ethanol. But, in some cases, ethanol may be preferred because of its renewable nature if derived from agricultural resources.

From the transesterification process, unreacted methanol is considerably easier to recover than ethanol. Ethanol forms an azeotrope with water; hence, it is expensive to purify during recovery. If the water is not removed, it will interfere with the reactions. On the other hand, methanol recycles easier because it does not form an azeotrope with water. Because of these reasons, methanol is the preferred alcohol, despite its high toxicity (i.e., exposure to methanol can cause blindness and other ill effects).

Table 7.9. *Inputs and mass requirements for the Lurgi process to produce 1 ton of biodiesel (Lurgi, 2009)*

Input	Requirement
Vegetable oil	1,000 kg
Methanol	96 kg
Catalyst	5 kg
Hydrochloric acid (37%)	10 kg
Caustic soda (50%)	1.5 kg
Steam requirement	415 kg
Electricity	12 kWh
Nitrogen	1 Nm^3
Process water	20 kg

Dry methanol is very corrosive to some aluminum alloys, which can be completely inhibited by the addition of 1% water. However, it must be noted that methanol with too much water (>2%) becomes corrosive again. Commercial ethanol typically contains some acetic acid and is particularly corrosive to aluminum alloys.

7.5.2 Batch Process

The simplest method for producing biodiesel is to use a batch stirred-tank reactor. Alcohol-to-triglyceride mole ratios from 4:1 to 20:1 have been reported, with a 6:1 being the most common. The reactor may be sealed or equipped with a reflux condenser, operating at about 67°C, although temperatures from 25°C to 85°C have been reported (Ma and Hanna, 1999; Demirbas, 2002a; Bala, 2005). The most commonly used catalyst is sodium hydroxide with a loading of 0.3–1.5%. Typical reaction times range from 20 minutes to longer than 1 hour. About 85–95% transesterification is achieved, although a higher temperature and a higher alcohol-to-oil ratio can enhance the percent completion.

The oil is first charged to the reactor, followed by the catalyst and methanol. The mixture is agitated during the reaction time. After the reaction, agitation is stopped, and the reaction mixture is allowed to settle in the reactor for separation of the esters and glycerol. Alternatively, the reaction mixture is pumped into a settling vessel or is separated using a centrifuge (Van et al., 2004).

The alcohol is removed from both the glycerol and the ester streams using an evaporator or a flash separator. The esters are neutralized by gently washing with warm, slightly acid water to remove residual methanol and salts, and then drying. The finished biodiesel is then sent to storage. Similarly, the glycerol stream is neutralized and washed with soft water, and then sent to the glycerol refining unit. In a two-step Lurgi process, most of the glycerin is recovered after the first stage, when a rectifying column separates the excess methanol and crude glycerin. The methyl ester output from the second stage is purified to some extent of residual glycerin and methanol by a wash column. Table 7.9 shows the inputs and mass requirements for the Lurgi process for illustration of input and utility needs.

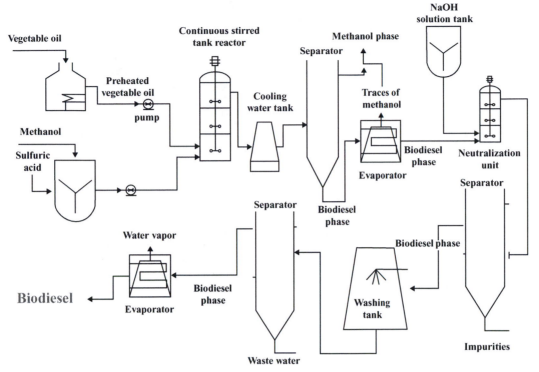

Figure 7.6. A schematic diagram of a continuous unit for biodiesel production from vegetable oil using an economical process (adapted from Chongkhong et al., 2009).

7.5.3 Continuous Process

A popular variation of the batch process is the use of continuous stirred-tank reactors in a series. Intense mixing using pumps or motionless mixers is used to initiate the esterification. A typical process scheme using a sulfuric acid catalyst is shown in Figure 7.6 (Chongkhong, Tongurai, and Chetpattananondh, 2009). In this process, optimum conditions for palm oil feedstock were found to be 8.8:1:0.05 molar ratio for methanol:oil:sulfuric acid, for 60 minutes of reaction time at 75°C. For larger scale operations, the continuous operation provides lower production costs compared with the batch operation.

7.5.4 Single-Phase Cosolvent Process

The low solubility of alcohol in the oil phase is attributed to slow esterification. In addition, free fatty acids present in the oils react with the base catalyst. These problems were addressed by the single-phase cosolvent process developed by Boocock et al. (1996, 2001). The process is commercialized by Biox Corporation (Figure 7.7). In this process, a cosolvent (e.g., tetrahydrofuran) is used to mix alcohol and oil into one phase. The single-phase reaction of alcohol and oil is much faster than the duel-phase system. In the first step, an acid catalyst is used to convert free fatty acids. In the second step, base is added to catalyze transesterification of triglycerides. After

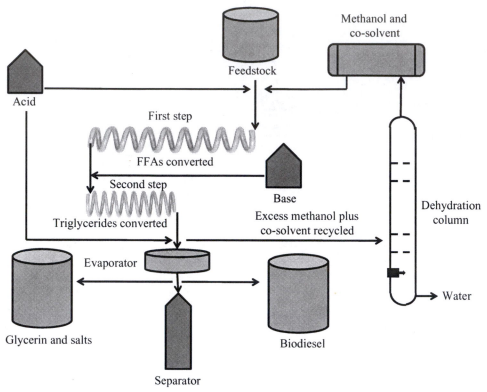

Figure 7.7. Biox process (adapted from Biox Corporation, 2009).

completion of transesterification, an evaporator is used to evaporate off the cosolvent and unreacted methanol, which are then recycled. The residue glycerol and biodiesel automatically phase separate. A high alcohol-to-oil ratio (typically 15–35:1) is used, which allows for over 99% conversion in a short time. In addition, the cosolvent process can easily handle a variety of feedstocks, including vegetable oil, animal fats, and waste cooking greases (Van et al., 2004).

7.5.5 Supercritical Methanol Process

Current technologies have difficulty with transesterification of low-quality feedstocks (e.g., waste cooking and industrial oils), as the free fatty acid contained reacts with alkali catalyst to form soap. The noncatalytical supercritical methanol process addresses these challenges, as first proposed by Kusdiana and Saka (2001) and Demirbas (2002a). Recently, Kusdiana and Saka (2004a) have proposed a two-step supercritical process that uses milder reaction conditions. The scheme (Figure 7.8) follows a two-step reaction: hydrolysis and esterification. Subcritical water is used to hydrolyze triglycerides into fatty acids; the glycerol byproduct leaves with water and is later separated. The oil phase, containing mostly fatty acids, is taken for methanol transesterification in the second reactor. The unreacted methanol is recycled back,

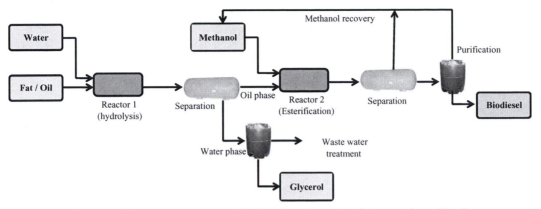

Figure 7.8. Scheme of the two-step supercritical methanol method (adapted from Kusdiana et al., 2004a).

and the biodiesel product is taken for purification. The two-step process completes the transesterification in a shorter reaction time under milder reaction conditions than the direct supercritical methanol process. Typical time needed for hydrolysis is 270 minutes and for methanol esterification is 20 minutes, in the case of rapeseed oil feedstock.

7.6 Fuel Properties of Biodiesels

Fuel properties of biodiesels are characterized by their viscosity, density, cetane number, cloud and pour points, distillation range, flash point, ash content, sulfur content, carbon residue, acid value, copper corrosion, and, most importantly, the higher heating value. The produced biodiesel should meet the requirements set by the standards. Biodiesel standard, ASTM D-6751, specifies the quality for pure biodiesel (B100). The international standard, EN 14214, describes the minimum requirements for biodiesel produced from rapeseed. These standards are shown in Tables 7.10 and 7.11.

7.6.1 Viscosity, Density, and Flash Point

The high viscosity of pure vegetable oil is the major problem associated with its direct use in diesel engines. Viscosity is reduced by transesterification of vegetable oil (Demirbas, 2003). For example, viscosity of vegetable oil is reduced from 27.2–53.6 mm^2/s to 3.6–4.6 mm^2/s upon transesterification, which is inline with the viscosity of No. 2 diesel fuel (2.7 mm^2/s at 40°C). Viscosity is an important property of biodiesel because it impacts the operation of fuel injection equipment, especially at low temperatures when the rise in viscosity affects the fluidity of the fuel. High viscosity leads to a poor atomization of the fuel spray, which results in a less accurate operation of the fuel injectors. The lower the viscosity of the biodiesel, the easier it is to pump and atomize and achieve finer droplets for combustion. The flash point of biodiesel is significantly lower than that of vegetable oil, which aids in the

Table 7.10. *North American biodiesel, B100, specifications (ASTM D-6751–02)*

Property	Method	Limits
Flash point	D-93	130°C min
Water and sediment	D-2709	0.050 vol% max
Kinematic viscosity at 40°C	D-445	1.9–6.0 mm²/s
Sulfated ash	D-874	0.020 wt% max
Total sulfur	D-5453	0.05 wt% max
Copper strip corrosion	D-130	No. 3 max
Cetane number	D-613	47 min
Cloud point	D-2500	
Carbon residue	D-4530	0.050 wt% max
Acid number	D-664	0.80 mg KOH/g max
Free glycerin	D-6584	0.020 wt% max
Total glycerin	D-6584	0.240 wt% max
Phosphorus	D-4951	0.0010 wt% max
Vacuum distillation end point	D-1160	360°C max, at 90% distilled

Note: An updated version, D-6751–09, has been published by ASTM (2009).

Table 7.11. *European biodiesel specifications (EN-14214)*

Property	Units	Lower limit	Upper limit	Test method
Ester content	% (m/m)	96.5	–	Pr EN 14103d
Density at 15°C	kg/m³	860	900	EN ISO 3675/ EN ISO 12185
Viscosity at 40°C	mm²/s	3.5	5.0	EN ISO 3104
Flash point	°C	>101	–	ISO CD 3679e
Sulfur content	mg/kg	–	10	–
Tar remnant (at 10% distillation remnant)	% (m/m)	–	0.3	EN ISO 10370
Cetane number	–	51.0	–	EN ISO 5165
Sulfated ash content	% (m/m)	–	0.02	ISO 3987
Water content	mg/kg	–	500	EN ISO 12937
Total contamination	mg/kg	–	24	EN 12662
Copper band corrosion (3 h at 50°C)	rating	Class 1	Class 1	EN ISO 2160
Oxidation stability, 110°C	hours	6	–	pr EN 14112k
Acid value	mg KOH/g	–	0.5	pr EN 14104
Iodine value	–	–	120	pr EN 14111
Linoleic acid methyl ester	% (m/m)	–	12	pr EN 14103d
Polyunsaturated (≥4 double bonds) methyl ester	% (m/m)	–	1	–
Methanol content	% (m/m)	–	0.2	pr EN 141101
Monoglyceride content	% (m/m)	–	0.8	pr EN 14105m
Diglyceride content	% (m/m)	–	0.2	pr EN 14105m
Triglyceride content	% (m/m)	–	0.2	pr EN 14105m
Free glycerine	% (m/m)	–	0.02	pr EN 14105m/ pr EN 14106
Total glycerine	% (m/m)	–	0.25	pr EN 14105m
Alkali metals (Na + K)	mg/kg	–	5	pr EN 14108/ pr EN 14109
Phosphorus content	mg/kg	–	10	pr EN14107p

Table 7.12. *Viscosity, density, and flash point measurements of nine oil methyl esters (Demirbas, 2008a)*

Methyl ester of	Viscosity mm^2/s (at 40°C)	Density kg/m^3 (at 15°C)	Flash point (°C)
Cottonseed oil	3.8	870	160
Hazelnut kernel oil	3.6	860	149
Linseed oil	3.4	887	174
Mustard oil	4.1	885	168
Palm oil	3.9	880	158
Rapeseed oil	4.6	894	180
Safflower oil	4.0	880	167
Soybean oil	4.1	885	168
Sunflower oil	4.2	880	166

combustion. Some of the properties of biodiesels derived from various oils are listed in Table 7.12.

7.6.2 Cetane Number, Cloud, and Pour Point

The cetane number (CN) is a measure of ignition quality of diesel fuels; a high CN implies short ignition delay, allowing more time for combustion to complete. High-speed diesel engines operate more effectively with high CN fuels. The scale is based on two compounds: hexadecane with CN of 100 and heptamethylnonane with CN of 15. The CNs of biodiesels (Table 7.13) are generally higher than conventional diesel (CN values in the range of 40–45). The longer the fatty acid carbon chains and the more saturated the molecules, the higher the CN. (Bala, 2005).

For operation in cold climates, cloud point (CP) and pour point (PP) of the fuel are important. CP is the temperature at which wax first becomes visible (i.e., cloudy look) when fuel is cooled, and PP is the temperature at which a sufficient amount of wax precipitates out of solution to gel the fuel to stop the flow. Hence, lower values of both CP and PP are desired for operation in cold climates. Biodiesels have both higher CP and PP compared with conventional diesel.

Table 7.13. *Cetane numbers of six methyl ester biodiesels (Demirbas, 2008a)*

Biodiesel source	Centane number
Sunflower	49
Soybean	46
Palm	62
Peanut	54
Babassu	63
Tallow	58

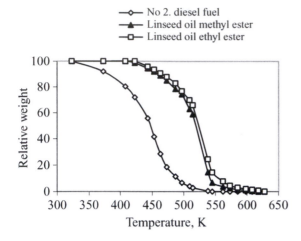

Figure 7.9. Distillation curves for No. 2 diesel fuel and linseed oil methyl and ethyl esters.

7.6.3 Combustion Efficiency

The oxygen content of biodiesel improves the combustion efficiency, especially due to the fact that oxygen atoms are part of the the ester molecules. Biodiesel contains about 11 wt% oxygen. The homogeneity of oxygen in the fuel is a great plus for biodiesel, which provides higher combustion efficiency than petrodiesel. Visual inspection of the injectors shows no difference between the biodiesel fuels and when tested on petrodiesel. In addition, the use of biodiesel can extend the life of diesel engines because it is more lubricating than petroleum diesel fuel.

7.6.4 Comparison of Methyl with Ethyl Esters

In general, the fuel properties of ethyl esters are comparable to those of the methyl esters. Both types of esters have almost the same energy content. However, the viscosities of the ethyl esters are slightly higher, and the CP and PP are slightly lower than those of the methyl esters. Engine tests demonstrated that methyl esters produce slightly higher power and torque than ethyl esters (Encinar et al., 2002). Some positive attributes of the ethyl esters over methyl esters include significantly lower smoke opacity, lower exhaust temperatures, and lower PP. However, ethyl esters tend to have more injector coking than methyl esters. The volatility of methyl esters is higher than that of ethyl esters, but both esters are less volatile than No. 2 diesel (Figure 7.9).

7.6.5 Emissions

Replacement of diesel with biodiesel can provide a substantial decrease in the amount of exhaust emissions. For example, a 90% reduction in total unburned hydrocarbon and a 75–90% decrease in polycyclic aromatic hydrocarbons have been reported. In addition, biodiesel provides a significant reduction in particulates and carbon monoxide emission when compared with petroleum diesel. However,

Table 7.14. *Biodegradability data of petroleum and biofuels*

Fuel	Degradation in 28 days	References
Gasoline (91 octane)	28%	Speidel, Lightner, and Ahmed, 2000
Heavy fuel (Bunker C oil)	11%	Mulkins-Phillips and Stewart, 1974; Walker, Petrakis, and Colwell, 1976
Refined rapeseed oil	78%	Zhang et al., 1998
Refined soybeen oil	76%	Zhang et al., 1998
Rapeseed oil methyl ester	88%	Zhang et al., 1998
Sunflower seed oil methyl ester	90%	Zhang et al., 1998

biodiesel provides a slight increase or decrease in nitrogen oxides, depending on engine family and testing procedures. Sulfur content of petrodiesel is 20–50 times that of biodiesels; hence, sulfur dioxide emission from biodiesel is very small. Also, the use of biodiesel can reduce net carbon dioxide (CO_2) emission. For example, the reductions in net CO_2 emissions are estimated at 77–104 g/MJ of diesel displaced by biodiesel.

7.6.6 Biodegradability

Biodiesel is nontoxic and, due to its oxygen contents, degrades about four times faster than petrodiesel. The degradation of biodiesel is compared with petroleum and vegetable oils in Table 7.14. Spills of petroleum liquids have been a problem for the ecosystem. In the case of biodiesel, less severe problems are expected, as chemicals from biodegradation of biodiesel can be released into the environment. Additional challenges arise from the small-scale processing plants that do not dispose/treat their methanol- and glycerol-containing wastes properly. With the increasing interest in biodiesel, health and safety aspects are of utmost importance, including environmental impacts of transport, storage, and processing (Ma and Hanna, 1999).

7.6.7 Engine Performance

On a volumetric basis, biodiesel has about 10% less energy content compared with diesel, mainly stemming from the oxygen in biodiesel. But the same oxygen content is responsible for a higher combustion efficiency of biodiesel, which partly compensates for the lower energy content. Hence, biodiesel provides, on average, 5% less power compared with diesel at a given load. In general, for biodiesel, the torque curves are flatter, and the peak torque is less than that for diesel and occurs at a lower engine speed.

Certain types of elastomers and natural rubbers (primarily fuel hoses and fuel pump seals) can soften and degrade as a result of contact with biodiesel over time. However, this effect is lessened when using biodiesel/diesel blends. For high-biodiesel-containing blends, biodiesel-compatible elastomers such as Viton B should be used. In fact, with the recent switch to low-sulfur diesel, most automobile manufacturers have switched to compatible elastomers. Also, carbon deposits inside the engine, with the exception of intake valve deposits, and the engine wear rates are normal with biodiesel.

Biodiesel provides a significant improvement in lubricity over diesel. For example, even blending 1% biodiesel can provide up to a 30% increase in lubricity (Demirbas, 2008a). This is a significant benefit, as fuel injectors and some types of fuel pumps rely on fuel for lubrication. Also, fuel lubricity is important for reducing friction wear in engine components that are normally lubricated by the fuel rather than crankcase oil (Ma and Hanna, 1999; Demirbas, 2003).

7.7 Disadvantages of Biodiesel

Compared with diesel, the major disadvantages of biodiesel are higher viscosity, lower energy content, higher CP and PP, higher emissions of nitrogen oxides (NO_x), lower engine speed and power, injector coking, engine compatibility, and higher price. In addition, the important operational disadvantages are cold start problems, higher copper strip corrosion, and pumping difficulty due to the higher viscosity.

A significant disadvantage with biodiesel is that it competes with food supply of vegetable oil. Additional competition with the food grains occurs when farmers switch from grain crops to oil seed crops. A solution is to grow oil plants on land that is not being used for growing food. The food/fuel competition will likely keep biodiesel prices high, due to the increasing demands for food worldwide.

7.8 Summary

Concerns about CO_2 emission and depeletion of petroleum diesel make biodiesel an attractive alternative motor fuel for compression ignition engines. Biodiesel, a renewable and biodegradable fuel, is derived from a variety of oils and fats by transesterification with ethanol or methanol. The transesterification process is influenced by the molar ratio of triglycerides (oils or fats) to alcohol, catalysts (acid or base), reaction temperature, reaction time, and free fatty acids and water content of triglycerides. In the conventional transesterification process, free fatty acids and water always produce negative effects; the presence of free fatty acids and water causes soap formation, reduces catalyst effectiveness, and consumes catalyst, which results in a low conversion. To solve these problems, a noncatalytic supercritical alcohol process has been developed. Biodiesel is more expensive than diesel, largely due to the high price of vegetable oil feedstock. Biodiesel can be successfully used in diesel automobiles, with only a slight decrease in power and minor changes in the elastomers.

8 Diesel from Biomass Gasification Followed by Fischer–Tropsch Synthesis

8.1 Diesel Fuel

The diesel engine was conceptualized by Rudolph Diesel in 1893; in this engine, air is compressed, resulting in a rise in temperature. The design eliminated the need for an external ignition source, and the compression heat was enough to cause combustion with fuel. The early use of this engine was a stationary application to run heavy machinery. In the 1920s, the engine was redesigned into a smaller size that was suitable for the automobile industry. At the 1911 World's Fair in Paris, Dr. Diesel ran his engine on peanut oil and declared "the diesel engine can be fed with vegetable oils and will help considerably in the development of the agriculture of the countries which use it" (Nitschke and Wilson, 1965). The use of vegetable oil and transesterified vegetable oil continued for some time, and later the engine was reengineered to run only on petroleum fuel. Since then, diesel consumption has grown significantly (Figure 8.1), putting pressure on the world supply and environment.

8.1.1 Diesel from Petroleum

Currently, the main source of diesel is petroleum crude oil. The crude oil is separated by distillation into various fractions, which are further treated (e.g., cracking, reforming, alkylation, polymerization, and isomerization treatments) to produce saleable products. Figure 8.2 shows a schematic of a typical refining process and various products.

Crude oil is typically heated to 350–400°C and piped into the distillation column kept at atmospheric pressure. A temperature gradient of 20–400°C is maintained in the column, in which vapor rises and liquid falls to the bottom, passing through a series of perforated trays. The lighter hydrocarbons remain in the vapor for longer and ultimately condense on the top trays; similarly, heavier hydrocarbons condense more quickly and reach the bottom trays. In this way, the light gases (methane, ethane, propane, and butane) are collected from the top of the column, petrol from the top trays, kerosene from upper-middle trays, diesel from middle trays, fuel oils from bottom trays, and heavy residue from the bottom of the

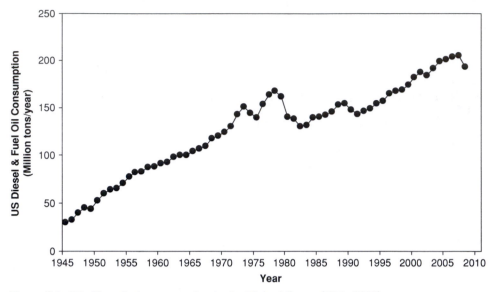

Figure 8.1. Distillate fuel consumption in the United States (EIA, 2009).

column. To recover additional heavy distillates (lubricating oils, waxes, and bitumen), the residue is sent to a second distillation column maintained under vacuum. The second column allows heavy hydrocarbons with boiling points of 450°C and higher to be separated without partly cracking into low-value products, such as coke and gas. The recovered heavy distillate is further refined using solvent extraction to produce lubricating oil.

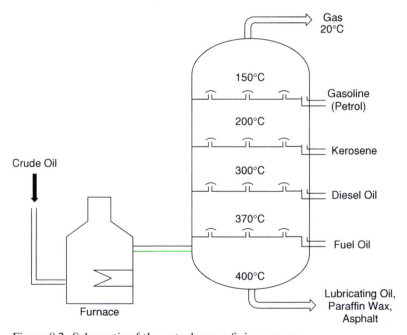

Figure 8.2. Schematic of the petroleum-refining process.

Table 8.1. *Typical composition of coal (wt% basis)*

Coal	Water	Ash	C	H	O	N	S
Lignite	15.2	7.0	54.1	3.9	19.0	0.8	1.1
Sub-bituminous	4.8	7.6	67.7	5.8	13.0	1.1	0.2
Bituminous	4.6	6.4	72.3	4.2	11.6	0.9	0.4

Petroleum diesel, which is simply referred to as diesel, is the fraction that boils in the range of 200–350°C at atmospheric pressure. Diesel is composed of ~75% saturated hydrocarbons (primarily n-, iso-, and cyclo-paraffins) and 25% aromatic hydrocarbons (naphthalenes and alkylbenzenes). The formula of molecules ranges from $C_{10}H_{20}$ to $C_{15}H_{28}$, with an average of $C_{12}H_{23}$. Diesel is sold in various grades, with a quality characterized by sulfur content, aromatic content, ignition quality, cold weather properties, content of pollutants, viscosity, density, and boiling point. Additional properties (e.g., centane number) and standards are discussed in Chapter 7.

Traditionally, diesel contained a high amount of sulfur. Off-road vehicles may use diesel containing more than 500 ppm sulfur, but on-road vehicles have been using low-sulfur diesel (up to 500 ppm). Starting in 2007, most places require the use of ultra-low-sulfur diesel that contains less than 15 ppm sulfur. Hence, petroleum diesel requires desulfurization to meet the standards. Lowering sulfur content reduces the lubricity of the fuel, which requires the use of lubricating additives.

8.1.2 Diesel from Coal

To obtain diesel from coal, first coal is gasified to synthesis gas (or syngas), a mixture of carbon monoxide (CO) and hydrogen (H_2), and then cleaned to remove impurities. The clean syngas is sent to a Fischer–Tropsch (FT) synthesis reactor, in which diesel and other liquids are produced using a catalyst. During the gasification, steam is used to provide needed hydrogen as

$$C + H_2O \rightarrow H_2 + CO.$$

Depending on the source and type, coal contains a good amount of sulfur (Table 8.1), which needs to be removed before taking syngas for FT synthesis. Some catalysts require sulfur concentration to be as low as 60 ppb; on the other hand, a sulfided catalyst works fine with a sulfur range of 50–100 ppm.

A variety of coal gasification technologies are available (Lin, 2009), including: (1) EAGLE Gasification Technology, which uses entrained upflow with a two-stage swirling burner in a single chamber; (2) Texaco (GE) technology, which employs entrained downflow with a slurry feed similar to a natural gas or heavy residual oil gasifier. Slurry of coal powder and water is fed into the gasifier from the top using a slurry pump. Typical operating temperature is 1330–1500°C with a residence

Table 8.2. *Gas product from gasification of coal using a Shell Gasifier (Lin, 2009)*

Gas	Volume %
H_2	26.7
CO	63.3
CO_2	1.5
H_2O	2.0
H_2S	1.3
N_2	4.1
Ar	1.1

time of a few seconds; and (3) Shell Gasification Technology, which is based on the entrained upflow, oxygen-blown, slagging gasification using dry-feed system. Here, coal powder is injected into the gasifier using nitrogen gas. The raw gas is cleaned by filtering and water scrubbing to remove fly ash, and by acid gas-treating including sulfur recovery. Typical composition of gas from a Shell Gasifier is shown in Table 8.2.

Liquid fuel is synthesized from syngas as

$$nCO + (n+1)H_2 \rightarrow (CH_2)_n + H_2O$$

using a FT catalyst described later.

8.1.3 Diesel from Biomass

In FT synthesis of diesel, coal can be replaced by biomass, a renewable energy source (Jin and Datye, 2000; Tijmensen et al., 2002). A typical scheme for biomass-derived diesel and other fuels is shown in Figure 8.3. The three key steps in the conversion technology are: (1) gasification of the biomass, (2) cleaning of the syngas, and (3) usage of the syngas to produce liquid fuel via FT synthesis.

8.2 Gasification of Biomass to Produce Syngas

Although biomass gasification processes vary considerably, typical operation is at $\geq 700°C$ and 1–5 atmosphere pressure. The process is generally optimized for the particular feedstock (e.g., wood, straw, grasses, coconut shells, and rice husks) and to properly handle the byproducts char, ash, and slag. In most processes, biomass is introduced at the top of the reactor, and the gasifying medium is either introduced co-currently (downdraft) or counter-currently up through the packed bed (updraft). Additional designs incorporate circulating or bubbling fluidized beds. Due to the relative simplicity of the gasifiers, operators experienced with conventional boilers and furnaces are able to operate the system. The modular design allows for a wide range of scales, for example, 1–20 tons/hour of biomass, depending on the availability of the feedstock.

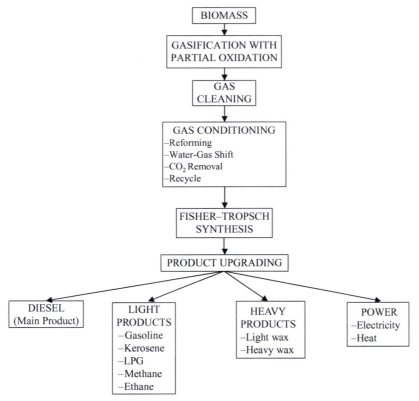

Figure 8.3. Diesel and other products from biomass via FT synthesis.

The key steps in the gasification process (Figure 8.4) are

1. Delivery of biomass to the dryer;
2. Reforming in the gasifier producing hot syngas, which contains ash and small amount of unreformed char;
3. Recovery of heat from hot syngas; and
4. Cleaning of syngas.

Depending on the particle size and consistency, biomass is often routed through a hammer-mill or tub grinder/classifier before entering the gasifier's metering bin located above the feed system. Biomass is fed by gravity into the metering bin, where it enters a screw feed system. The biomass is then conveyed through a sealing mechanism that serves as the pressure seal on the front end of the system, keeping air out of the gasifier and keeping syngas from backing up into the feed system. The biomass received from the screw feeder is conveyed with recycled compressed syngas into the primary heat exchanger. In the primary heat exchanger, biomass is preheated, devolatilized, and dried, and the hot syngas is cooled to the desired temperature suitable for gas filter. The preheated, partially reformed/gasified biomass and conveying syngas pass from the convection section of the primary heat exchanger into the radiant coil section of the primary reformer, where high-temperature steam

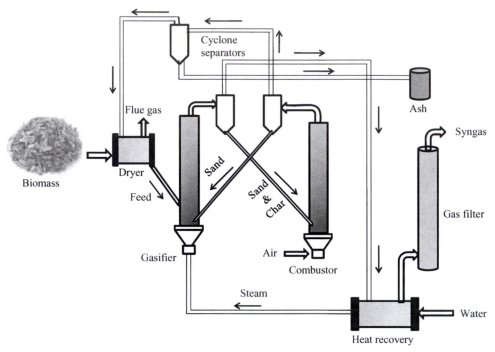

Figure 8.4. Battelle biomass gasification system.

reforming takes place. The syngas and any char (ash and any unreformed carbon) are taken to a barrier-type gas filter element, where the char is removed as a dry residue. This air-cooled heat exchanger receives clean syngas from the gas filter and reduces gas temperature to the desired level for supplying power generation equipment or other fuel uses.

8.2.1 Types of Gasifiers

For use in many industrial applications, gasifiers with various configurations have been developed, including

1. Fixed-bed (updraft or downdraft fixed beds) gasifiers,
2. Fluidized-bed (fluidized or entrained solids serve as the bed material) gasifiers, and
3. Others, including moving-grate beds and molten salt reactors.

Schematics of updraft (counter-current) and downdraft (co-current) fixed-bed gasifiers are shown in Figure 8.5.

In the updraft case, biomass is fed at the top and air is fed from the bottom via a grate (Figure 8.5a). Char falls toward the grate, and in the zone immediately above the grate, it is combusted with the temperature reaching $\sim 1000°C$. Ash, generated after the combustion, falls through the grate. The produced hot gas rises upwards and is reduced. Rising further up, gas pyrolizes the biomass and, in the top zone,

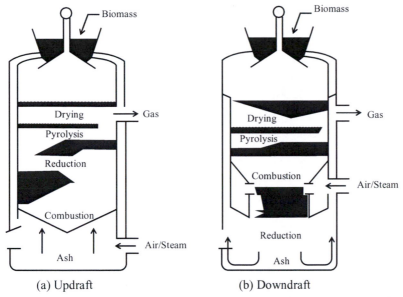

Figure 8.5. Schematic of updraft (a) and downdraft (v) biomass gasifiers (adapted from Mc-Kendry, 2002).

dries the biomass, all this resulting in the cooling of the gas to about 200–300°C. In the pyrolysis zone, a significant amount of tar is formed, some of which leaves the gasifier with the product gas and the remaining condenses on the biomass higher up. The biomass feed provides a filtering effect that captures the particles, producing a gas in low particulate content. The temperature in the reduction zone is controlled by adding the steam to the air used for gasification. As the heat of leaving gas is recovered by the biomass, the overall energy efficiency of the process is high; however, the tar content is also high.

In the downdraft gasifier, the air and the biomass move in the same direction (Figure 8.5b). Because the product gas leaves the gasifier right after passing through the hot zone, the tar is partially cracked, giving low tar content in the product gas. But the product gas carries a high amount of particulate content. The overall energy efficiency is low because the gas leaves at about 900–1000°C and high-energy content is carried along with the hot gas.

In the cross-flow gasifier, the air is introduced from the side and the biomass moves downward. The product gas is withdrawn opposite of the air inlet at the same level. A pyrolysis/drying zone is formed higher up in the gasifer, and a combustion/gasification zone is formed near the air feed. The product gas leaves the gasifier at about 800–900°C, and the ash is removed at the bottom. The overall energy efficiency of the cross-flow gasifier is low and the tar content of the gas is high.

Fluidized-bed gasifiers provide a much more uniform temperature in the gasification zone as compared with the fixed-bed gasifiers. A bed of fine-grained material is used in which air is introduced, and the fluidization ensures the intimate mixing of the bed material, combustion gas, and the biomass feed. There are two types of fluidized beds in use: circulating bed and bubbling bed. Circulating fluidized bed

Table 8.3. *Gasification reactions (McKendry, 2002)*

	Reaction	Heat of reaction
Partial oxidation	$C + 0.5\,O_2 \leftrightarrow CO$	-268 MJ/kg-mole
Complete oxidation	$C + O_2 \leftrightarrow CO_2$	-406 MJ/kg-mole
Water–gas reaction	$C + H_2O \leftrightarrow CO + H_2$	$+118$ MJ/kg-mole
Water–gas-shift reaction	$CO + H_2O \leftrightarrow CO_2 + H_2$	-42 MJ/kg-mole
Methane formation	$CO + 3H_2 \leftrightarrow CH_4 + H_2O$	-88 MJ/kg-mole

can be designed for high throughput, especially for gasification of bark and forestry residue in the pulp and paper industry. The bed material is circulated between the main gasifier vessel and a cyclone separator where ash is removed, and the char and bed material are returned to the gasification vessel. The bubbling bed gasifer consists of a reactor with a grate at the bottom on which a bed of fine-grained material is kept. The air is introduced from the bottom of the grate with enough velocity to bubble through the bed material. The biomass is fed on the bed material, which is maintained at 700–900°C by adjusting the air-to-biomass ratio. In the hot bed, biomass is pyrolyzed to form char and gaseous compounds. The product has low tar content (McKendry, 2002).

8.2.2 Gasification Chemistry

Gasification, a complex thermochemical process, proceeds with a number of elementary chemical reactions, beginning with the partial oxidation of biomass with the gasifying agent (usually air, oxygen, and/or steam). The chemical reactions involved can be categorized as follows: flash evaporation of inherent moisture, devolatization of heavy organics, cracking of heavy hydrocarbons, pyroylysis, and steam reforming.

Most gasifiers use air or oxygen for partial oxidation of volatile matter to yield the combustion products water (H_2O) and carbon dioxide (CO_2), plus heat to continue the endothermic gasification of biomass. Product gas is composed of CO, CO_2, H_2O, H_2, methane (CH_4), other gaseous hydrocarbons (CHs), tars, char, inorganic components, and ash. The gas composition can vary greatly, depending on the gasification process, the gasifying agent, and the biomass composition. A generalized reaction scheme can be written as:

$$\text{Biomass} + O_2 \rightarrow CO, CO_2, H_2O, H_2, CH_4 + \text{other(CHs)} + \text{tar} + \text{char} + \text{ash}$$

$$(8.1)$$

With key individual reactions as shown in Table 8.3.

The relative amounts of CO, CO_2, H_2O, H_2, and CH_x depend on the stoichiometry of the gasification. When air is used as the gasifying agent, roughly half of the product gas is nitrogen (N_2). The air-to-biomass mass ratio in a gasification process is generally kept in the range of 0.2–0.35. And if steam is the gasifying agent, the steam-to-biomass mass ratio is around 1.

Table 8.4. *Heating value of syngas using various gasification agents (McKendry, 2002)*

Gasification agent	Heating value of gas	Remark
Air and air/steam	4–6 MJ/Mm^3	Low heating value
O_2 and steam	12–18 MJ/Mm^3	Medium heating value
H_2	40 MJ/Mm^3	High heating value

For optimum gas yield, char gasification is important as it is the slowest reaction. Studies have shown that char gasification is almost a zero-order reaction with the reaction rate being fairly constant throughout and declines only when the char is nearly depleted. The reaction rate depends on the surface area of the char particles and proximity to any catalyst (e.g., sodium or potassium salts). Due to the catalytic effect, the mineral (ash) content of the original biomass can influence the char gasification (Demirbas, 2000b).

The usefulness of syngas to produce diesel is directly linked to the heating value of the gas. Various gasification agents can give rise to differing heating values of the product gas, as shown in Table 8.4. The overall biomass-to-syngas conversion efficiency based on energy is about 75–80%. Also the gasification efficiencies based on carbon can be achieved as high as 97%. For efficient use of carbon in FT synthesis, it is important to minimize CH_4 formation during gasification and to convert all carbon in the biomass to CO and CO_2 (Prins, Ptasinski, and Janssen, 2004).

8.3 Conditioning of Syngas

The biomass-derived syngas contains needed FT building blocks CO and H_2. The H_2/CO mole ratio in the syngas from air gasification is ≤ 1, which is not well suited for synthesis of FT diesel. On the other hand, steam gasification of biomass provides an H_2/CO mole ratio of >1, which is suitable for synthesis of liquid hydrocarbons. One can tune the ratio by using a mixture of gasifying agents and a water–gas shift reaction. Ideally, a good mole ratio is 2:1. To maximize the use of the carbon sources, the steam reforming of biosyngas with additional natural gas and other feedstock can be considered (Bukur et al., 1995; Larson and Jin, 1999; Byrd, Pant, and Gupta, 2007, 2008). The effect of gas composition on the yield of various components is shown in Figure 8.6.

Syngas also contains undesired impurities, including tar, hydrogen sulfide, carbonyl sulfide, ammonia, hydrogen cyanide, alkali, and dust particles. These impurities must be thoroughly removed to avoid poisoning of the FT catalyst (Stelmachowski and Nowicki, 2003). Impurities from syngas can be removed by several methods, including physical separation, thermal cracking, and catalytic hot gas cleanup (Yung, Jablonski, and Magrini-Bair, 2009). In physical separation technologies, such as filtration and aqueous/organic liquid scrubbing, the gas is first cooled to near-ambient temperature, and the impurities are captured in the filter, and/or scrubbing liquids, which need to be regenerated. The cleaned gas is heated back

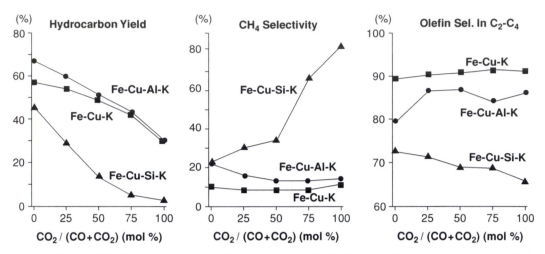

Figure 8.6. Dependence of catalytic activity and selectivity on feed gas composition (balanced feed gas: $H_2/(2CO + 3CO_2) = 1$, 1 MPa, 300°C, gas-hourly-space velocity = 1,800 mL/ (gcat h) (Jun et al., 2004; reproduced with permission of Elsevier).

to the FT temperature. Thermal cracking is done at a high temperature (>1100°C), which can remove most of the tar present in the syngas. Additionally, catalytic cleaning can also condition the syngas (Yung et al., 2009), without much heating/cooling thermal penalties or generation of waste streams.

8.4 FT Synthesis to Produce Diesel

FT synthesis is a catalytic process for producing mainly straight-chain hydrocarbons from CO and H_2 contained in syngas. FT synthesis is typically carried out in the temperature range of 210–340°C and at high pressures (15–40 bar). The product range includes light hydrocarbons (CH_4 and C_2H_6), propane (C_3H_8), butane (C_4H_{10}), gasoline (C_5–C_{12}), diesel (C_{13}–C_{22}), and waxes (C_{23}–C_{33}). The relative distribution of the products depends on the catalyst and the process conditions (temperature, pressure, and residence time). Due to the catalyst used, the feed syngas must have very low tar and particulate matter content (Jun et al., 2004; Demirbas, 2007b).

The FT synthesis, originally established in 1923 by German scientists Franz Fischer and Hans Tropsch, is described by a set of equations as

$$nCO + (2n + 1)H_2 \rightarrow C_nH_{2n+2} + nH_2O \qquad (8.2)$$

where n is the average the number of carbon atoms in the hydrocarbon produced (Schulz, 1999). All reactions (for varying value of n) are exothermic ($\Delta H = -165$ kJ/mol CO), and the product is a mixture of different hydrocarbons mainly consisting of paraffins and olefins. The product distribution is described by Anderson–Schulz–Flory as

$$x_n = (1 - \alpha)\alpha^{n-1} \qquad (8.3)$$

where x_n is the mole fraction of the hydrocarbon with n carbon atoms, and α is the chain growth probability (Patzlaff et al., 1999).

The value of parameter α is influenced by the composition of the syngas, temperature, pressure, and the composition of the catalyst. Since value of α is less than unity, FT synthesis favors the formation of methane ($n = 1$), and the mole fraction decreases as the n increases (i.e., for product molecules containing more carbon atoms).

A main characteristic regarding the performance of FT synthesis is the selectivity toward liquid products (Stelmachowski and Nowicki, 2003). For the above reaction (Equation 8.2), an H_2/CO ratio of at least 2 is required for the synthesis of hydrocarbons. When iron (Fe)-based catalysts, with water–gas shift reaction activity, are used, the water produced in the reaction (Equation 8.2) can react with CO to form additional H_2, as

$$CO + H_2O \rightarrow CO_2 + H_2 \quad \Delta H = -42 \text{ kJ/mol} \tag{8.4}$$

In this case, a minimal H_2/CO ratio of 0.7 is required, as

$$2nCO + (n + 1)H_2 \rightarrow C_nH_{2n+2} + nCO_2 \quad \Delta H = -102 \text{ kJ/mol CO} \tag{8.5}$$

Fe catalysts are inexpensive, have a high tolerance for sulfur, and produce olefin- and alcohol-rich products. But the lifetime of the Fe catalysts is short, usually limited to 8 weeks in commercial installations (Davis, 2002). Research is still ongoing for the development of a bulk Fe catalyst that combines high FT activity, low methane selectivity, long-term stability, and high attrition resistance. It appears that the critical property determining the activity and deactivation of Fe catalyst is not to be Fe in the metallic state but the carburized Fe surface. The bulk Fe catalysts are the catalysts of choice for converting low H_2/CO ratio syngas obtained from biomass.

8.4.1 Reactor Configurations

The FT synthesis plant consists of FT reactors, recycle and compression of unconverted syngas, removal of H_2 and CO_2, reforming of CH_4 produced, and separation of the products. The most important aspects for commercial FT synthesis are the integration of high-reaction heats and the large number of products (gas, liquid, and solid hydrocarbons) with varying vapor pressures. The main categories of reactors are fixed bed, slurry bed, and fluidized bed, as shown in Figure 8.7. The first two categories are suitable for diesel production, and the third category is suitable for gasoline and olefin production. The main reactor types that have been proposed and developed after 1950 are (1) multitubular fixed-bed reactor with internal cooling; (2) circulating fluidized-bed reactor with circulating solids, gas recycle, and cooling in the gas/solid recirculation loop (Synthol; Sasol, South Africa); (3) fluidized-bed reactors with internal cooling (SAS; Sasol); and (4) three-phase fluidized (ebulliating)-bed reactors or slurry bubble column reactors with internal cooling tubes (Steynberg et al., 2004). The fixed-bed reactor consists of thousands

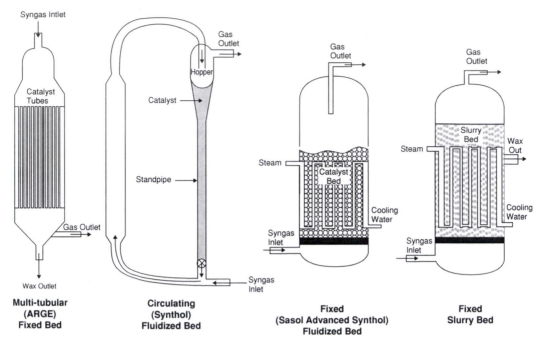

Figure 8.7. Types of FT synthesis reactors (Spath and Dayton, 2003).

of small tubes containing catalyst. Water surrounds the tubes and regulates the temperature. In the slurry reactor, the catalyst has no particularly position but flows around with the reaction components. In all of the designs, heat removal and temperature control are the most important design features to obtain optimum product selectivity and long catalyst lifetimes.

Key considerations in FT reactor selection design are type of desired FT products, source and composition of syngas, process conditions (temperature, pressure, gas space–velocity, etc.), functional characteristics and scalability of reactors, catalyst composition and physical characteristics, catalyst deactivation, catalyst loading, physical properties of the reaction medium (viscosity, density, thermal conductivity, surface tension, reactants diffusivity, mass- and heat-transfer coefficients, dispersion coefficients, etc.), and layout of the reactor's internal components (Guettel, Kunz, and Turek, 2008).

Present commercial FT reactors operate at two different temperature ranges: (1) the high-temperature FT (HTFT) operates at about 340°C with Fe-based catalysts mainly for the production of gasoline and olefins, and (2) the low-temperature FT (LTFT) operates at about 230°C with either Fe- or cobalt-based catalysts for the production of diesel and linear waxes. Although the process is designed to maximize one type of product, the FT always produces a range of product, from methane to waxes. In addition to paraffins and olefins, oxygenated products, such as alcohols, aldehydes, and carboxylic acids, are also produced. At HTFT conditions, aromatics and ketones are also produced. Typically, slurry- and fixed-bed reactors are used in LTFT, and the fluidized-bed reactor is used in HTFT.

For exothermic reactions, multitubular reactors are widely used, as they are easy to handle and design and have uniform tube behavior. Typical tube diameters are a few centimeters and catalysts dimension are 1–3 mm. The disadvantages include high capital cost, limited catalyst loading for limiting pressure drop, possibility of hot spot formation damaging the catalyst, and need for a significantly low feed and cooling medium temperature. Slurry bubble reactor for FT synthesis uses bubble column with 10- to 200-µm size suspended catalyst. The small catalyst particle size provides negligible mass-transfer resistance, giving an optimum activity and selectivity. However, the separation of the fine catalyst particles from the liquid products is a challenge on a large scale. Another development in the reactor design is the use of microstructured reactors. Similar to honeycombs, a large number of parallel small channels can give rise to excellent mass- and heat-transfer characteristics. Even for highly exothermic reactions, isothermal operation can be achieved. This approach appears to be promising for decentralized and mobile applications. It is claimed that the high conversion and low methane selectivity can reduce the capital cost (Guettel et al., 2008).

8.4.2 Catalysts

FT synthesis is defined as hydrogenation of CO on metallic catalysts yielding long carbon-chain hydrocarbons (Van Steen and Claeys, 2008). In addition to metals, metal carbide and metal nitride can also catalyze FT synthesis. CO molecules adsorb on the catalyst and undergo dissociation due to the strong interaction. The adsorbed carbon atoms join to form chains, which then combine with H_2 atoms to produce hydrocarbons. For FT synthesis, catalytic activity is of utmost important in the running of the plant. Two main classes of catalyst used are Fe-based and cobalt-based catalysts. Fe-based catalysts have a high tolerance for sulfur, produce more olefin and alcohols, are inexpensive, and have a short lifespan (about 8 weeks in commercial installations). On the other hand cobalt-based catalysts have a high conversion rate, a longer life (over 5 years), and generally produce less unsaturated hydrocarbons and alcohols. Additional catalysts are based on nickel and ruthenium. Nickel was used in early FT work, was found to produce too much methane at low pressure, and had too much loss as nickel carbonyl at high pressures. Hence, the use of nickel has been abandoned. And the high price of ruthenium has kept its use very limited (Table 8.5).

8.4.2.1 Iron-Based Catalysts

Fe catalysts (typically in the form of unsupported Fe/Cu/K prepared by precipitation) have successfully been used for many decades due to their key advantages: (1) low cost compared with cobalt or ruthenium; (2) high water–gas-shift activity allowing use of low H_2-containing syngas from biomass; (3) high activity for the production of liquids and waxes, which are readily refined to diesel and gasoline; and (4) high selectivity for C_2–C_6 olefin hydrocarbons, which can be used as chemical feedstocks. There are also notable disadvantages: (1) high loss due to attrition;

Table 8.5. *Relative prices of metals*
(Van Steen and Claeys, 2008)

Metal	Relative price[a]
Iron	1
Copper	32
Nickel	140
Cobalt	235
Silver	2,100
Palladium	49,000
Ruthenium	76,000
Gold	114,000
Platinum	203,000
Rhodium	824,000

[a] Price per kg in 2007, relative to 1 kg of scrap iron.

and (2) irreversible deactivation over a period of a few months to a few years. The catalyst deactivation is due to sintering, oxidation, formation of inactive surface carbons, and carbide phase transformation (Jothimurugesan et al., 2000; Dry, 2002; Jun et al., 2004; Wu et al., 2004).

Fe is the most widely used catalyst by Sasol for FT synthesis (Dry, 2002). Precipitated Fe is used in LTFT reactors with multitubular fixed-bed or slurry reactor configuration operating at 220–250°C and 25 bar. Fused iron catalyst is used in HTFT reactors with fixed (SAS) or circulating (Synthol) configuration operating at 300–320°C and 25 bar. Typically, Fe-based catalysts contain a significant amount of magnetite, which renders a fraction of Fe within the catalysts inactive (Van Steen and Claeys, 2008). If the fraction of carbide is increased, the productivity of the catalysts should increase. Thus, it has been suggested that increasing crystallite size may result in more Fe being active. With Fe catalyst, copper is added to enable easy reduction, and potassium is added as a promoter (Riedel et al., 1999).

8.4.2.2 Cobalt-Based Catalysts

Cobalt-based catalysts are preferred when high productivity, long catalyst life, selectivity for linear hydrocarbons, and low-activity for competing water–gas shift reaction are needed (Jean-Marie et al., 2009). Both activity and selectivity of supported cobalt catalysts depend on the number of cobalt surface atoms, atom density within support particles, and transport limitations that restrict access to these sites. Altering the catalyst preparation method can affect these factors. The variables in catalyst preparation include cobalt precursor type and loading level, support structure and composition, amount of promoters, and the pretreatment procedure. Acid sites can be introduced on the catalyst, which can induce the formation of branched paraffins directly during the FT synthesis. However, water can inhibit these secondary isomerization reactions on the acid sites, and also oxidize cobalt sites, making them inactive for additional turnovers (reactions). This is of special concern because water can be produced in large amounts; one water molecule is produced for each carbon atom

Table 8.6. *Effects of Al_2O_3/SiO_2 ratio on hydrocarbon selectivity (Jothimurugesan et al., 2000)*

Hydrocarbon selectivities (wt%)	100Fe/ 6Cu/5K/ 25SiO$_2$	100Fe/6Cu/ 5K/3Al$_2$O$_3$/ 22SiO$_2$	100Fe/6Cu/ 5K/5Al$_2$O$_3$/ 20SiO$_2$	100Fe/6Cu/ 5K/7Al$_2$O$_3$/ 18SiO$_2$	100Fe/6Cu/ 5K/10Al$_2$O$_3$/ 15SiO$_2$	100Fe/ 6Cu/5K/ 25Al$_2$O$_3$
CH$_4$	6.3	8.7	10.4	10.7	14.3	17.3
C$_{2-4}$	24.5	27.8	30.8	29.9	33.4	46.5
C$_{5-11}$	26.8	27.6	32.2	33.9	40.0	31.0
C$_{12-18}$	21.9	21.2	15.8	15.0	6.0	4.9
C$_{19+}$	20.5	14.4	11.0	10.6	6.1	0.4

Reaction condition: 523 K, 2.0 MPa, and $H_2/CO = 2.0$, and gas stream velocity: 2,000 h^{-1}.

added to the growing hydrocarbon chain, and due to the low water–gas-shift activity of cobalt.

When clean syngas is available, cobalt-based catalysts are the preferred catalysts (Van Steen and Claeys, 2008). However, process upset in the syngas cleaning section may deactivate the catalyst, which is expensive to replace. On the other hand, such process upsets are of less severe financial burden in the case of Fe-based catalysts, which are more resistant to sulfur and ammonia poisoning.

8.4.2.3 Catalyst Supports

The catalyst supports used in FT synthesis include alumina, silica, and titania (Al_2O_3, SiO_2, and TiO_2). Support is of key importance in the case of expensive metal catalysts, when it is important to avoid crystallite sintering. The support itself has effect on the catalytic activity; for example, the selectivity of low-molecular-weight hydrocarbons and paraffin-to-olefin ratio increases with increasing Al_2O_3/SiO_2 ratio for Fe catalyst as shown in Table 8.6 (Jothimurugesan et al., 2000). Al_2O_3 as a structural promoter improves dispersion of copper and potassium, which gives a much higher FT synthesis activity (Jun et al., 2004).

8.4.3 FT Synthesis in Supercritical Fluids

The fixed-bed FT reactors are susceptible to local heating due to the highly exothermic nature of the reactions and condensation of waxes onto the catalyst pore due to poor removal of the heavy products. In gas-phase FT, reaction rate and product mass transfer are high, but heat removal is slow, resulting in high CH$_4$ formation. In the slurry reactors, some of these shortcomings are overcome with increased bulk heat and mass transfer; however, mass transfer in the catalyst micropores is still limited given overall slow reactions. An ideal medium for FT would have gas-like transport properties and liquid-like heat capacity, thermal conductivity, and solubility properties (Abbaslou et al., 2009). These requirements have led researchers to look into the application of supercritical fluids and compressed gas as reaction media.

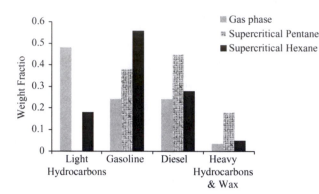

Figure 8.8. Distribution of productions from FT synthesis in gas phase, supercritical hexane, and supercritical pentane media (adapted from Elbashir and Roberts, 2004).

Supercritical fluids (SC) offer several advantages for heterogeneous and non-thermal reactions. Benefits derived from the SC Fischer–Tropsch synthesis (SC-FT) include gas-like diffusivities and liquid-like solubilities, which together combine the desirable features of the gas- and liquid-phase FT synthesis routes (Huang and Roberts, 2003). In addition, the reaction environment can be manipulated by small changes in pressure and/or temperature to enhance the solubility of the reactants and products and to eliminate phase-transfer limitation, and even to integrate separation process with the reactor.

SC such as CO_2, 1-pentane, and 1-hexane have been tested. The rapid desorption and dissolution of heavy hydrocarbon products from the catalyst sites creates more vacant sites in SC-FT; hence, the conversion of CO in SC medium is faster than in gas-phase medium. In addition, the overall product distribution in SC-FT shifts to heavier products as compared with gas phase (Figure 8.8). The olefin content in the product from SC-FT is higher than other FT reactors because the primary products (i.e., olefins) can be effectively removed from the catalysts sites and transported away by the SC before they are readsorbed and hydrogenated to paraffins (Abbaslou et al., 2009). Also, the uniformity of the temperature in SC-FT enhances the stability of the catalyst.

From the reactor operation and the high quality of the product, it is clear that SC-FT is superior to other configurations. However, capital and operating costs are likely to be higher for SC-FT. In addition, the large amount of added fluid in SC-FT must be separated and recycled. An economic evaluation regarding the viability of SC-FT is necessary. The evaluation should include both increased capital costs due to the high-pressure equipment and operating costs due to the compression of the SC for recycling. Also, the optimum pressure, temperature, and degree of dilution need to be established.

8.5 Fuel Properties of FT Diesel

After performing FT synthesis, the product is distilled into various fuel fractions and waxes. Also, the reactor can be configured to make product rich in the desired fuel. FT diesel is similar to petroleum diesel with regard to its density, viscosity, and energy content (Table 8.7). In fact, FT diesel is superior due to a high cetane

Table 8.7. *Comparison of FT diesel with No. 2 petroleum diesel*

Property	FT diesel	No. 2 petroleum diesel
Density	0.78 g/cm^3	0.83 g/cm^3
Higher heating value	47.1 MJ/kg	46.2 MJ/kg
Aromatics	0–0.1%	8–16%
Cetane number	76–80	40–44
Sulfur content	0–0.1 ppm	25–125 ppm

number and a low aromatic content, which results in low particle and nitric oxides (NO_x) emissions. FT diesel has excellent autoignition characteristics, as it is composed of only straight-chain hydrocarbons and has no aromatics or sulfur. FT diesel can be blended with petroleum diesel in any proportion without the need for infrastructure or engine modifications. The emissions (CO, NO_x, and particles) from FT diesel are much lower than those from combustion of petroleum diesel. Hence, this high-quality diesel either can be used in the locations with very strict emission requirements or can be blended with low-quality petroleum diesel to bring the overall quality up to meet the standards. The negative attributes of FT diesel include poor lubricity (but responds well to lubricant additives), poor cold flow properties, reduced elastomer swelling, and susceptibility to oxidation.

8.6 Summary

Biomass can be gasified to produce syngas, which can be used to produce diesel by FT synthesis. Various gasification technologies (updraft, downdraft, etc.) are available to produce syngas. After clean up and conditioning of syngas, it is sent to FT reactor where predominantly linear hydrocarbons are synthesized, which are of particular interest in producing high-quality diesel. Various FT reactor configurations (fixed bed, slurry bed, fluidized bed, supercritical, etc.) are available. Two catalysts of choice are iron and cobalt. Compared with petroleum diesel fuel, FT diesel is comparable or even better due to higher cetane number, lower aromatic content, and low sulfur, and it results in lower emissions.

9 Bio-Oil from Biomass Pyrolysis

9.1 What Is Bio-Oil?

Biomass, when subjected to high temperature in the absence of oxygen (i.e., pyrolysis), converts into gas, solid char, and liquid products. The liquid product, called bio-oil or pyrolysis oil, is typically brown, dark red, or black in color with a density of about 1.2 kg/liter. Bio-oil has a water content of typically 14–33 wt%, which cannot be easily removed by conventional methods (e.g., distillation); in fact, bio-oil may phase separate above certain water content. The higher heating value of bio-oil is typically 15–22 MJ/kg, which is lower than that for conventional fuel oil (43–46 MJ/kg), mainly due to the presence of oxygenated compounds in bio-oil (Mohan, Pittman, and Steele, 2006).

9.2 Pyrolysis

Pyrolysis is the thermal decomposition of biomass, which occurs in the absence of oxygen or when significantly less oxygen is supplied than needed for complete combustion. Pyrolysis can convert biomass into more useful fuels: a hydrocarbon-rich gas mixture, an oil-like liquid, and a carbon-rich solid residue.

Pyrolysis, in the form of wood distillation, has been used since historical times. For example, ancient Egyptians practiced wood distillation by collecting tars and pyroligneous acid for use in their embalming industry and for caulking boats. In the 1800s, wood pyrolysis to produce charcoal was a major industry for supplying fuel for the industrial revolution, until it was replaced by coal. Until the early twentieth century, wood distillation was still a profitable industry for producing soluble tar, pitch, creosote oil, chemicals, and noncondensable gases often used to heat boilers. The industry declined in the 1930s due to the advent of the lower-price petrochemical products.

The pyrolysis process is similar to gasification, except it is generally optimized for the production of bio-oil that can be used straight or refined for higher quality uses, such as engine fuels, chemicals, adhesives, and so on. Pyrolysis of biomass is somewhat different from that of coal, as (1) pyrolysis of biomass starts at a lower

Table 9.1. *Pyrolysis methods and their variants (Demirbas, 2005a; Mohan et al., 2006)*

Method	Residence time	Temperature (°C)	Heating rate	Products
Carbonation	days	400	very low	charcoal
Conventional	5–30 min	600	low	oil, gas, char
Fast	0.5–5 s	650	very high	bio-oil
Flash-liquid[a]	<1 s	<650	high	bio-oil
Flash-gas[b]	<1 s	<650	high	chemicals, gas
Hydropyrolysis[c]	<10 s	<500	high	bio-oil
Methanopyrolysis[d]	<10 s	>700	high	chemicals
Ultra pyrolysis[e]	<0.5 s	1,000	very high	chemicals, gas
Vacuum pyrolysis	2–30 s	400	medium	bio-oil

[a] Flash-liquid: Liquid obtained from flash pyrolysis accomplished in <1 s.
[b] Flash-gas: Gaseous material obtained from flash pyrolysis in <1 s.
[c] Hydropyrolysis: Pyrolysis with water.
[d] Methanopyrolysis: Pyrolysis with methanol.
[e] Ultra pyrolysis: Pyrolysis with very high degradation rate.

temperature; (2) volatile matter content in biomass is higher; (3) char from pyrolysis of biomass has more oxygen; and (4) ash from biomass is more alkaline in nature, which may aggravate fouling problems. Biomass pyrolysis is typically carried out in the temperature range of 400–700°C and at near-ambient pressure. The thermal degradation of lignocellulosic compounds follows a complex mechanism, which can be influenced by heating and cooling rates. For example, rapid heating and rapid quenching can produce intermediate bio-oil, which condenses before further decomposing into gaseous products. High reaction rate can minimize char formation; in fact, under some conditions, no char is formed. The main pyrolysis variants are listed in Table 9.1.

In broad terms, pyrolysis variants can be categorized into two classes: slow pyrolysis (or conventional pyrolysis, or carbonization) and fast pyrolysis; general differences between the two are listed in Table 9.2.

9.2.1 Slow Pyrolysis

Conventional or slow pyrolysis is defined as the pyrolysis that occurs under a slow heating rate. This technique has been used for thousands of years mainly to produce charcoal. The process conditions permit the production of all three solid,

Table 9.2. *Range of the main operating parameters for pyrolysis processes (Demirbas and Arin, 2002)*

	Slow pyrolysis	Fast pyrolysis
Temperature (°C)	275–675	575–975
Heating rate (°C/s)	0.1–1	10–200
Particle size (mm)	5–50	<1
Solid residence time (s)	450–550	0.5–10

liquid, and gaseous products in significant portions. During the first stage (prepy-rolysis stage) of biomass decomposition, which occurs at 120–200°C, some internal molecular rearrangement, such as water elimination, bond breakage, appearance of free radicals, and the formation of carbonyl, carboxyl, and hydroperoxide groups, takes place. During the second stage, the main decomposition of solid biomass takes place, which proceeds with a high rate and leads to the formation of the pyrolysis products. During the third stage, the char decomposes at a very slow rate, forming a carbon-rich residual solid. The heating rate in conventional pyrolysis is much slower than that in fast pyrolysis; in fact, biomass can even be held at a constant temper-ature for some time. Conventional pyrolysis is carried out when a high amount of charcoal is needed in the product.

9.2.2 Fast Pyrolysis

Fast pyrolysis (also called thermolysis) is a process in which biomass is rapidly heated to high temperatures in the absence of oxygen. To obtain a fast heating rate, high operating temperatures, very short residence times, and very fine particles are required. As a result, biomass decomposes to generate mostly vapors and aerosols and some charcoal. A high bio-oil production requires very low vapor residence time of typically 1 second to minimize secondary reactions; although, reasonable yields can be obtained at residence times of up to 5 seconds if the vapor temperature is kept below 400°C. After cooling and condensation, a dark brown liquid is formed that has a heating value about half that of conventional fuel oil. In contrast to slow pyrolysis, fast pyrolysis is an advanced process, which needs to be carefully con-trolled to produce high yields of bio-oil. A typical reaction temperature of around 500°C is employed. In fact, fast pyrolysis has now been adopted for the production of food flavors to replace the traditional slow pyrolysis, which had much lower yields. These processes use very short vapor residence times of between 30 and 1500 mil-liseconds and reactor temperatures around 500°C. Control of both temperature and residence time is important to "freeze" the intermediates of most chemical interest in conjunction with moderate gas/vapor phase temperatures of 400–500°C before recovery of the product to maximize organic liquid yields (Bridgwater, Meier, and Radlein, 1999).

Recent studies on fast pyrolysis have shown that high yields of primary, nonequilibrium liquids and gases, including valuable chemicals, chemical interme-diates, and bio-oil fuels, can be obtained from biomass. Thus, the higher-value fuel gas, bio-oil, or chemicals from fast pyrolysis can replace the lower-value solid char from traditional slow pyrolysis. The essential features of a fast pyrolysis process are (1) rapid heating of biomass, which usually requires a finely ground biomass feed; (2) carefully controlled pyrolysis reaction temperature of around 500°C in the vapor phase; (3) a short vapor residence time of typically less than 2 seconds; and (4) rapid cooling of the pyrolysis vapors to give a high bio-oil yield. Various design variables for fast pyrolysis include feed drying, particle size, pretreatment, reactor configuration, heat supply, heat transfer, heating rates, pyrolysis temperature, vapor

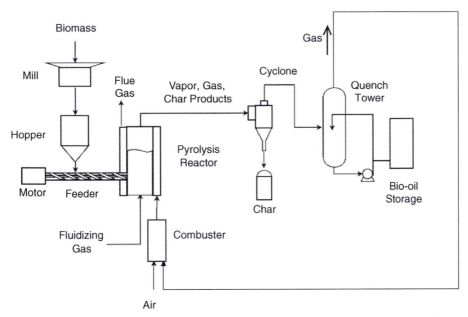

Figure 9.1. General process schemes for fast pyrolysis (adapted from Brown, 2003).

residence time, secondary cracking, char separation, ash separation, and liquid collection (Mohan et al., 2006).

Fast pyrolysis produces about 75% liquid, 12% char, and 13% gas, which is much better than the yields from slow pyrolysis of about 30% liquid, 35% char, and 35% gas, on mass basis. Due to the superiority of fast pyrolysis for bio-oil yield, the rest of this chapter focuses on fast pyrolysis.

9.3 Process Considerations

An overall pyrolysis scheme is shown in Figure 9.1. Pyrolysis is difficult to precisely define, especially when applied to heterogeneous materials such as biomass. The older literature generally equates pyrolysis to carbonization, in which the main product is a solid char. In contrast, today, the term pyrolysis often describes processes in which bio-oil is the preferred product. Fast pyrolysis, being much faster than conventional pyrolysis, has process steps that are controlled to a very small time scale. The general changes that occur during pyrolysis are enumerated below (Babu and Chaurasia, 2003; Mohan et al., 2006).

1. Heat transfer to increase the temperature of biomass;
2. The initiation of primary pyrolysis reactions and, at this higher temperature, volatiles are released and char is formed;
3. The flow of hot volatiles toward cooler biomass results in heat transfer from hot volatiles to cooler, unpyrolyzed biomass;
4. Condensation of some of the volatiles in the cooler parts of the biomass, followed by secondary reactions, which can produce tar;

5. Autocatalytic secondary pyrolysis reactions proceed while primary pyrolysis reactions (item 2, above) simultaneously occur in competition; and

6. Further thermal decomposition, reforming, water–gas-shift reactions, radical recombination, and dehydrations can also occur, depending on the residence time/temperature/pressure profile.

9.3.1 Feedstock Preparation

Bio-oil can be produced from a variety of dried forest residue and agricultural wastes. Biomass feedstock with good potential include bagasse (from sugarcane), rice straw, rice hulls, peanut hulls, oat hulls, switchgrass, wheat straw, coconut fibers, and wood. In North America and Europe, bio-oil is produced from forest residues, including sawdust, bark, and shavings. In other parts of the world, it is produced from sugarcane bagasse and other agricultural wastes. As received, biomass has a moisture content in the range of 50–60 wt%. Passive drying during summer storage can reduce the moisture content to 30 wt%, and active silo drying can reduce the moisture content to 12 wt%. Also, drying can be accomplished by using near-ambient solar or waste heat. Alternatively, specifically designed dryers can be used before pyrolysis (Mohan et al., 2006). The moisture content should be reduced below 10 wt%. In addition to drying, cleaning, washing, handling, grinding, storage, and transportation should be taken into account. For example, hot-water washing of high-ash biomass can improve oil quality and stability.

9.3.2 Heat Transfer Requirements

High-heat transfer rates are required for fast heating of the biomass particles. The heat required for pyrolysis is the total amount to provide for all the sensible, phase change, and reaction heats of the process. Only a small amount of reaction heat is needed, as the heat of reaction for the fast pyrolysis process is marginally endothermic. It is established that heat fluxes of 50 W/cm^2 are required to achieve true fast pyrolysis conditions. The majority of heat is transferred via convection and conduction modes, and the individual contribution from each mode will vary depending on the reactor configuration to give the most optimum heat transfer overall. In a circulating fluidized bed, the majority of the biomass heating comes from the hot circulating sand, typically at a sand-to-biomass ratio of 20:25. The large volume of sand use requires an effective sand-reheating system. On the other hand, in a conventional fluidized bed, sand requires an external heat supply, which can come from char burning. To obtain 62 wt% bio-oil yield from the prepared feedstock, the energy requirement, including radiation and exhaust gas losses, is about 2.5 MJ/kg of bio-oil produced. When noncondensable gases are directly used in the reactor burner, the energy requirement is only 1.0 MJ/kg of bio-oil produced, which may come from external fuels such as natural gas (Mohan et al., 2006).

9.3.3 Effect of Metal Ions and Salts

Several biomasses inherently contain metal ions and salts. Potassium and calcium are the major metal ions present, along with minor amounts of sodium, magnesium, and other elements. For others biomasses, metal ions can be added externally to alter or tailor the products from biomass pyrolysis. For example, the presence of alkaline cations is known to affect the mechanism of thermal decomposition. These cations cause fragmentation of the monomers from the natural polymer chains, rather than the predominant depolymerization that occurs in their absence. Due to the complex and heterogeneous nature of the system, these catalysts are not typically recycled back as in the case of conventional catalysts. Catalyzed biomass pyrolysis offers a great potential for future development as a route to modify bio-oil properties.

9.3.4 Catalysis

Bio-oil is not stable and is not miscible with conventional liquid fuels. Hence, it needs to be upgraded, especially to reduce its oxygen and water content. The use of catalyst during the pyrolysis can greatly improve the composition of bio-oil. A variety of catalysts including pure zeolite (HZSM-5), FCC catalysts, aluminas (α, γ-Al_2O_3), transition metal catalysts (Fe/Cr), and Al-MCM-41 have been tested (Goyal et al., 2007; Mohan et al., 2006). The bio-oil obtained by catalytic pyrolysis does not require costly preupgradation processing involving condensation and reevaporation.

9.3.5 Kinetics

Understanding the kinetics involved in pyrolysis can greatly help with the optimization and control of the process; however, the kinetics is complicated by the heterogeneous nature and biomass, for example, hemicelluloses, cellulose, lignin, and extractives each have their own unique pyrolysis chemistry. However, the simplified steps are the following: (1) degradation of biomass into primary products: tar, gas, and semi-char, (2) decomposition of primary tar and semi-tar to secondary products, and (3) continuous interaction between primary gas and char. Several modeling studies have been published (Authier et al., 2009; Bech et al., 2009; Papadikis, Gu, and Bridgwater, 2009) and some of these have disregarded the last step completely. Also, the effects of the biomass composite morphology on heat-transfer differences and the actual chemical degradation kinetics of individual biomass components have been modeled. The key parameters for the modeling of pyrolysis, bed temperature, particle size, feed rate, and especially reaction temperature, have a major important role in biomass pyrolysis. In addition, the accurate knowledge of thermal properties (including the specific heat) is necessary to model reactor kinetics effectively.

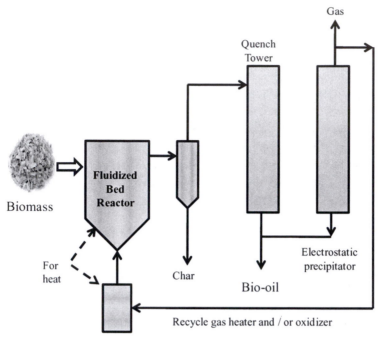

Figure 9.2. Bubbling fluidized-bed reactor (adapted from Bridgwater, 2003).

9.3.6 Bio-Oil Yields

Typical bio-oil yields are in the 60–80 wt% range, depending on the biomass composition (Mohan et al., 2006). For example, the bio-oil yields from wood are in the range of 72–80 wt%, depending on the relative content of cellulose and lignin in the feedstock. A lower bio-oil yield (60–65 wt%) is obtained from high lignin contents, such as that found in bark. However, the bio-oil obtained from a higher lignin-containing biomass does have a higher energy density/content. Mixed-paper feedstock gives a yield in excess of 75 wt%.

9.4 Pyrolysis Reactors

The key unit in the pyrolysis process is the pyrolysis reactor, for which considerable research and development has taken place over the last two decades. To meet the requirement of rapid heat transfer, various reactor designs have been explored. Some of these reactors have given bio-oil yield of as high as 70–80 wt% from dry biomass. The reactors can be categorized as (1) bubbling fluidized bed, (2) circulating fluidized bed, (3) vacuum pyrolysis, (4) ablative fast pyrolysis, (5) rotating cone pyrolyzer, and (6) auger type (Mohan et al., 2006).

9.4.1 Bubbling Fluidized Bed

The simple fluidized bed in which gas velocity is maintained to keep the bed in a suspended state is referred to as the bubbling fluidized bed (Figure 9.2). The

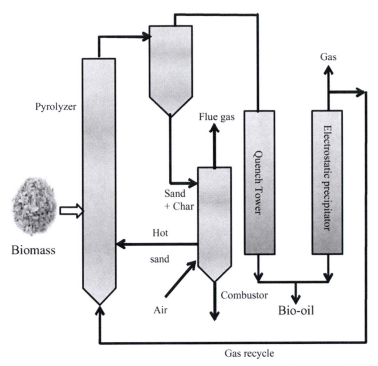

Figure 9.3. Circulating fluidized bed (adapted from Bridgwater, 2003).

construction and operation of such reactors is simple, and the fluidization technology involved is well understood. A high loading of solid bed (e.g., sand) is used that provides a uniform temperature and efficient heat transfer (Scott, 1988). The residence time of the biomass, char, and gas is controlled by flow rate of the fluidizing gas. Char, the heaviest component, has the longest residence time, but is eluted. The eluted char is further separated using cyclone separator(s), based on the difference in the aerodynamics of char and sand particles. While in the reactor, char does catalyze the cracking of heavy molecules in the vapor phase. Overall, the bubbling fluidized bed produces a good-quality bio-oil at a high yield.

9.4.2 Circulating Fluidized Bed

At a higher velocity of the fluidizing gas, all components – gas, bed (sand), and char particles – are carried out of the reactor to the cyclone separator. This led to the design of the circulating fluidized bed, where sand particles are circulated back to the reactor (Figure 9.3). The char particles have about the same residence time as gas, which leads to higher char content in the product bio-oil. The heat transfer primarily relies on the gas–solid convection, which is not particularly high when compared with solid–solid conduction in bubbling fluidized bed. The circulating bed is dilute in solid content. Commonly, a twin-bed reactor is used, in which the second vessel is used for combustion of char to reheat the circulating solids. The ash particles can build up in the reactor as some of them are carried back with the return circulation. These ash particles go on to catalyze the decomposition of the volatile

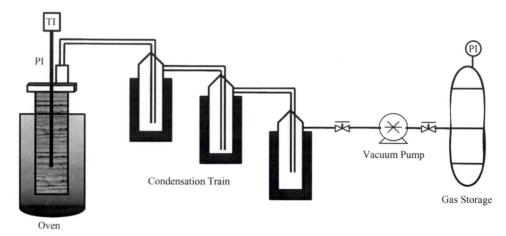

Figure 9.4. Schematic of the vacuum pyrolysis (adapted from Garcia-Perez, Chaala, and Roy, 2002).

components; hence, the bio-oil produced has low volatile content. Despite complex particle dynamics, the circulating fluidized bed has to be scaled up to handle very large throughputs.

9.4.3 Vacuum Pyrolysis

In vacuum pyrolysis, biomass is thermally decomposed under reduced pressure (Figure 9.4). Although the biomass is not heated rapidly, the gas generated is removed rapidly enough from the reactor, giving it a low residence time comparable to other fast-pyrolysis processes. The process is typically carried out at 450°C and at a total pressure of 15 kPa (i.e., 86 kPa below atmospheric pressure). Upon heating, the biomass decomposes into primary fragments (mostly high-boiling-point compounds), which are vaporized and rapidly removed from the reaction by vacuum. The secondary decomposition is avoided. Thus, the chemical/molecular composition of the bio-oil closely resembles the biomass feedstock. The short residence for volatiles is easily achieved by controlling the vacuum, and the residence time of volatiles is not linked with the residence time of the biomass particles, which stay for much longer in the reactor to complete the decomposition. However, the low pressure gives poor convective heat transfer. In addition, the vacuum pump and large reactor vessels are needed, increasing the capital cost.

9.4.4 Ablative Fast Pyrolysis

In the pyrolysis discussed so far, the processes are highly dependent on the heating rate of the biomass. The heating is achieved by gas-to-biomass convection or solid-bed particle-to-biomass conduction. The challenges for fast heat transfer are due to the gas-boundary layer in the case of convection and poor particle-to-particle contact in the case of conduction case. Typically, conduction heat transfer is faster than convection, but good surface contact is needed on the microscopic level. Therefore,

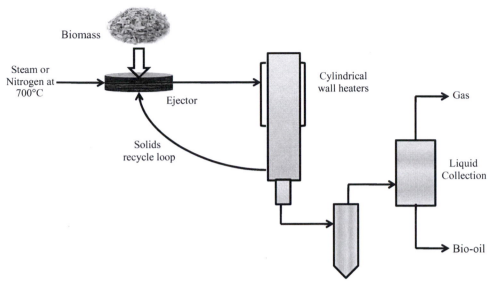

Figure 9.5. Schematic of the ablative bed reactor (adapted from Mohan et al., 2006; Bridgwater, 2003).

it is important to have small particle size or large surface area in fast pyrolysis processes. Ablative pyrolysis solves this problem by mechanically pressing the biomass against the heated reactor wall, which provides a much higher heating rate, inducing rapid melting and sublimation of biomass.

In the ablative fast pyrolysis process (Figure 9.5), biomass particles are injected onto the hot reactor wall, which causes pressing upon impact. Steam or nitrogen can be used to eject the biomass particles from the feeder onto the reactor wall. Additionally, centrifugal action can be used to push the particles toward the reactor wall. Reactor wall is typically kept at 600°C, and large biomass particles can be used. Due to the rapid heating, the reaction system is more intense. A high relative motion between the wall and the particle is used to continue to move the biomass particle forward with "scrapping" action. As the biomass is mechanically moved away, a residual oil film is created which both provides lubrication for successive biomass particles and rapidly evaporates to produce pyrolysis vapors. The vapors are condensed and collected similar to other pyrolysis schemes. The reaction rate is strongly dependent on the impact pressure, the relative velocity of biomass on the heat-exchange surface, shear forces that reduce particle size and increase surface area, and the reactor's wall temperature. The process is mechanically driven, so it is more complex. In addition, the process is surface area-driven, and therefore it is expensive to scale up.

9.4.5 Rotating Cone Pyrolyzer

In a rotating cone pyrolyzer, biomass particles are fed to the center of a rotating cone reactor (Janse et al., 1999). The centrifugal force causes the particles to move along the cone wall in a spiral way and pass the upper edge of the rotating cone.

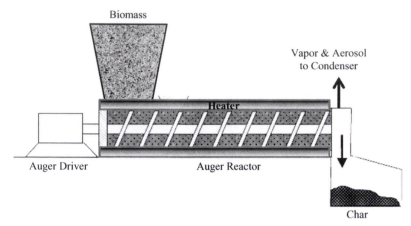

Figure 9.6. Schematic of the auger reactor (adapted from Brown and Holmgren, 2009).

The pyrolysis of the biomass particles starts immediately after entering the reactor. Sand is also added with the biomass particles, to avoid fouling of the cone wall and to increase the wall-to-particle heat transfer. This type of pyrolyzer is compact in design, operates at atmospheric pressure, and has a high biomass capacity. However, very small biomass particle size (50–100 µm) is required, and the process scale-up is difficult.

9.4.6 Auger Reactor

In this type of reactor, biomass particles are moved using an auger (a screw-type conveying) through a cylinder tube (Figure 9.6). The tube is heated externally or hot sand can be co-fed with the feed, causing biomass to devolatilize and gasify. Vapors are condensed as bio-oil, and the noncondensable portion is taken as bio-gas product. The vapor residence time can be modified by changing the heated zone of the tube, which in turn can allow control of the secondary pyrolysis reactions. Typically, temperature in the range of 400–800°C is used. The process operates continuously and does not require a carrier gas. The energy cost to run the Auger-type reactors is usually lower than other designs.

9.4.7 Future Developments

The designs of pyrolysis reactors are continuing to evolve. Success of the reactor is judged based on economically competitive performance, proven scale-up technology, and trouble-free operation. Unfortunately, none of the present designs fully satisfy all of the requirements. Hence, further development in the technology is needed with the following guiding elements.

1. The pyrolysis reactor should operate at the minimum possible temperature, giving an optimum bio-oil yield. If the temperature is kept below 450°C, many of the materials of construction issues become easier.

Table 9.3. *Properties of bio-oil (from wood pyrolysis) compared with fuel oils (Shaddix and Hardesty, 1999; Czernik and Bridgwater, 2004)*

Property	Bio-oil	Diesel fuel (No. 2)	Heavy fuel oil (No. 6)
Elemental composition (wt%)			
C	54–58	87.3	87.7
H	5.5–7.0	12.0	10.3
O	35–40	0	1.2
N	0–0.2	<0.01	0.5
S	0–0.04	0.1	0.8
Ash	0–0.2	<0.001	0.07
Higher heating value (MJ/kg)	16–19	46.7	41
Density (kg/liter)	1.2	0.86	0.94
Water content	15–30		0.1
pH	2.5		
Viscosity at 50°C (mm^2/s)	11–115	2.6	180
Solids (wt%)	0.2–1.0		1
Distillation residue (wt%)	up to 50		1

2. An independent heat supply allows easier control and more flexible operation as the feedstock varies from time to time.

3. A small gas-to-biomass ratio should be used to reduce the bio-oil vapor-quenching requirements.

4. To avoid the formation of microparticulate carbon (soot) from decomposition of organic vapor and carbon monoxide, the temperature of the recycled gas stream should be kept below 600°C.

5. To avoid char build up in the case of a fluidized reactor, a "blow through" mode of operation should be retained to elute char particles.

6. The attrition of char particles should be minimized.

7. Char should be collected separately, to allow flexible use of this product.

9.5 Fuel Properties of Bio-Oil

Bio-oil differs significantly from petroleum fuels due to its very high viscosity, moisture content, and oxygen content, and a lower heating value (Table 9.3). The viscosity is comparable to heavy fuel oil, which depends on the biomass feedstock, pyrolysis temperature profile, degree of thermal degradation, degree of catalytic cracking, and the content of light components and water. The high water content (15–30 wt%) of bio-oil cannot be easily removed by conventional methods, such as distillation. Also, the high water content is responsible for low energy density, low flame temperature, and ignition difficulties. Addition of more water to bio-oil can lead to phase separation. In contrast to diesel and gasoline, which are nonpolar and do not absorb water, bio-oil is highly polar and can readily absorb water up to 35 wt% (Demirbas, 2007c). The higher heating value of bio-oil is 16–19 MJ/kg compared with 40–45 MJ/kg for conventional petroleum fuels.

9.5.1 Chemical Composition

The major compounds in bio-oil are derived form the decomposition of biomass components: hemicelluloses, cellulose, and lignin. Upon heating, cellulose and hemicelluloses initially break into lower-molecular-weight compounds, forming an "activated cellulose," which decomposes by two competitive reactions: one forming volatiles (anhydrosugars) and the other forming char and gases. The lignin mainly produces char because it is not readily cleaved to light fragments. For example, a progressive increase in the pyrolysis temperature of the wood led to the release of the volatiles, thus forming a solid residue that is chemically different from the original material (Demirbas, 2000b).

The bio-oil formed at 450°C contains a high concentration of compounds, such as acetic acid, 1-hydroxy-2-butanone, 1-hydroxy-2-propanone, methanol, 2,6-dimethoxyphenol, 4-methyl-2,6-dimetoxyphenol and 2-cyclopenten-1-one, and so on, with a high percentage of methyl derivatives. As the temperature is increased, some of these compounds transform via hydrolysis. The formation of unsaturated compounds generally involves a variety of reaction pathways, such as dehydration, cyclization, Diels-Alder cycloaddition reactions, and ring rearrangement. For example, 2,5-hexandione can undergo cyclization under hydrothermal conditions to produce 3-methyl-2-cyclopenten-1-one with very high selectivity of up to 81% (Demirbas, 2008a). The influence of temperature on the composition of bio-oil has been examined by gas chromatographic analysis (Demirbas, 2007c). The bio-oils are composed of a range of cyclopentanone, methoxyphenol, acetic acid, methanol, acetone, furfural, phenol, formic acid, levoglucosan, guaiacol, and their alkylated phenol derivatives (Table 9.4).

If biomass is completely pyrolyzed, resulting products are about what would be expected by pyrolyzing the three major components separately. The hemicelluloses break down first, at temperatures of 195–255°C. Cellulose follows in the temperature range of 235–345°C, with lignin being the last component to pyrolyze at temperatures of 275–500°C. As a result, a wide spectrum of organic compounds is present in the bio-oil (Beaumont, 1985). For example, degradation of xylan yields eight main products: water, methanol, formic, acetic and propionic acids, 1-hydroxy-2-propanone, 1-hydroxy-2-butanone, and 2-furfuraldeyde. With an increase in temperature, the methoxyphenol concentration decreases, whereas that of phenols and alkylated phenols increases. The formation of both methoxyphenol and acetic acid is possibly a result of the Diels-Alder cycloaddition of a conjugated diene and unsaturated furanone or butyrolactone.

9.5.2 Viscosity

Viscosities of bio-oils vary greatly depending upon the biomass and temperature/time history of the pyrolysis process (Diebold, 2000). Viscosity also increases upon storage. Hence, there have been efforts to reduce and/or stabilize the viscosity. Removal of char can reduce the tertiary reaction and slow the increase in viscosity

Table 9.4. *Relative amount of main organic compounds in bio-oil from* Pterocarpus indicus *biomass (Luo, Wang, and Liao, 2004)*

Compound	Relative %
Furfural	9.06
Acetoxyacetone, 1-hydroxyl	1.21
Furfural, 5-methyl	1.82
Phenol	2.55
2-Cyclopentane-1-one, 3-methyl	1.58
Benzaldehyde, 2-hydroxyl	2.7
Phenol, 2-methyl	5.04
Phenol, 4-methyl	0.51
Phenol, 2-methoxyl	0.27
Phenol, 2,4-dimethyl	9.62
Phenol, 4-ethyl	2.18
Phenol, 2-methoxy-5-methyl	4.15
Phenol, 2-methoxy-4-methyl	0.55
Benzene, 1,2,4-trimethoxyl	3.8
Phenol, 2,6-dimethyl-4-(1-propenyl)	4.25
1,2-Benzenedicarboxylic acid, diisooctyl ester	1.8
2-Furanone	5.7
Levoglucosan	6.75
Phenol, 2,6-dimethoxy-4-propenyl	3.14
Furanone, 5-methyl	0.49
Acetophenone, 1-(4-hydroxy-3-methoxy)	2.94
Vanillin	6.35
Benzaldehyde, 3,5-dimethyl-4-hydroxyl	4.54
Cinnamic aldehyde, 3,5-demethoxy-4-hydroxyl	2.19

(aging). For example, 90°C-storage aging rate of char-filter poplar bio-oil is 2.5 cP/h, which is one-fifth that of nonfiltered oak bio-oil (Diebold and Czernik, 1997). Even with this aging rate, viscosity may double in 1 year. The aging can be further slowed by addition of various compounds, such as methanol, ethanol, acetone, and ethyl acetate, with a varying degree of success (Table 9.5).

Table 9.5. *Viscosity increase rate of char-filtered poplar bio-oil stored at 90°C (Diebold and Czernik, 1997)*

Additive	Viscosity increase rate (cP/h)
None	2.50
Ethyl acetate (10 wt%)	0.36
MiBK (5 wt%) + methanol (5 wt%)	0.25
Ethanol (10 wt%)	0.22
Methanol (5 wt%) + acetone (5 wt%)	0.20
Acetone (10 wt%)	0.19
Methanol (10 wt%)	0.14

9.5.3 Density

The density of pyrolysis liquid is very high at around 1.2 kg/liter. On the mass basis, the higher heating value of bio-oil is around 17 MJ/kg (at 25 wt% water content), which is around 38% of the petroleum liquids. But the volumetric energy density of bio-oil is around 20.5 MJ/liter compared with 33.5 MJ/liter for petroleum liquids, which gives bio-oil around 61% energy on volume basis.

9.5.4 Acidity

Bio-oils have a substantial amount of carboxylic acids (e.g., acetic and formic acids), which give a low pH of 2–3. The low pH (or acidity) that makes bio-oil corrosive is pronounced at higher temperatures. Hence, the material of construction for equipment handling bio-oil needs to withstand corrosiveness. In addition, the strong acidity of bio-oils makes them extremely unstable.

9.5.5 Water Content

Bio-oil has a content of water as high as 15–30 wt%, which emanates from the original moisture in the biomass and the product of dehydration reactions during the pyrolysis and storage. The presence of water lowers the heating value, resulting in a lower flame temperature. On the other hand, the presence of water enhances the fluidity and reduces the viscosity, which is needed for atomization during combustion in the engine.

9.5.6 Oxygen

Depending on the biomass and severity of the pyrolysis (temperature versus time profile), the oxygen content of bio-oil is around 35–40 wt%, which is distributed in >300 oxygenated compounds. The high oxygen content leads to lower heating value and immiscibility with hydrocarbons. These compounds can also undergo polymerization reactions during slow heating, giving rise to complex behavior during distillation. For example, bio-oils start to boil below 100°C and continue until 250–280°C, leaving 25–50 wt% residues at the end. Hence, bio-oils are not suitable for cases where complete evaporation is needed before combustion.

9.5.7 Char and Particle Content

The large char particles are separated out using cyclone separators, but some fine char particles are carried over and collected in bio-oil. Typically, the solid content in the bio-oil is around 1 wt%. Due to the small particle size, the surface area is high, which can catalyze various reactions during storage. In addition, the particle can cause blockage, erosion, and high emissions from incomplete combustion.

Two main methods are employed for removal of char particles: hot vapor filtration and liquid filtration. In hot vapor filtration, the gas is filtered right after cyclone separators and before cooling. This method has been successful, as the viscosity of the fluid is still low at high temperature. In liquid filtration, the bio-oil is filtered using cartridge filters, rotary pressure filters, or centrifuges. However, with this method, the filtration media is rapidly blocked due to the gel-like components from lignin pyrolysis (Bridgwater et al., 1999).

9.5.8 Storage Stability

Due to the rapid heating and cooling involved, bio-oil is the product of reactions that are far from reaching equilibrium. The produced bio-oil contains a large number of oxygenated compounds, typically in small percentages (Table 9.4). During the storage, the chemical composition of bio-oil changes toward thermodynamic equilibrium, resulting in an increase in the molecular weight and viscosity and decrease in the cosolubility of its many compounds. Due to these molecular transformations, the single-phase bio-oil can separate into various tarry, sludgy, waxy, and thin aqueous phases. Formed sludge and waxes, even in previously filtered bio-oil, can easily block the fuel filters.

9.6 Upgrading of Bio-Oil

The molecular weights of molecules in bio-oil range from 30 to 1000 g/mol, and boiling points range from 19 to 386°C (Branca and Di Blasi, 2006). When distilling pyrolysis oil at 150°C, up to 40 wt% mass forms residue, which creates handling difficulties (Deng et al., 2009). The polarity of compounds ranges widely, which results in a very heterogeneous mixture. For effective use as a fuel, bio-oil needs to be separated and concentrated into the compounds of similar polarities. And to increase the heating value for fuel use, the oxygen content must be lowered. To achieve these conditions, solvent fractionation and deoxygenation are practiced to upgrade the bio-oil.

9.6.1 Solvent Fractionation

Upon addition of water, bio-oil can be easily separated into an aqueous fraction and an organic fraction. Water-soluble fraction rich in carbohydrate-derived compounds collects as the top phase, and a viscous oligomeric lignin-containing fraction settles as the dense bottom phase. The large number of oxygenated compounds partition themselves into both phases. About 25–30% of the bio-oil goes into the lignin-rich fraction. Various usages are being proposed for the lignin-rich fraction, including the replacement of phenol formaldehyde resins and wood preservation to replace chromated copper arsenate and creosote.

Developments of commercially feasible bio-oil fractionation technique are still under development. Gavillan and Mattschei (1980) used stepwise addition of strong acids and bases to separate phenols, acids, and neutral compounds. And to avoid large waste volumes containing salts of bases and acids, Chum and Black (1990) used ethyl acetate solvent extraction. Recently, Deng et al. (2009) proposed the use

Table 9.6. *Properties of hydrotreated bio-oil as compared with gasoline (Holmgren et al., 2008)*

Component	Hydrotreated bio-oil (wt%)	Typical gasoline (wt%)
Paraffin	5.2–9.5	44.2
Iso-paraffin	16.7–24.9	35
Olefin	0.6–0.9	4.1
Naphthene	39.6–55.0	7
Aromatic	9.9–34.6	38
Oxygenate	0.8	

of glycerol for extraction due to its high boiling point (290°C) and low cost. Glycerol is a byproduct from the bio-diesel industry with increasing worldwide supply and decreasing prices. The extraction scheme is shown in Figure 9.7.

9.6.2 Deoxygenation

The negative properties (low heating value, high viscosity, storage instability, and corrosiveness) can be improved if oxygen content is reduced in the bio-oil. Therefore, an upgrading process for reducing oxygen content is required before the final use. The recent upgrading techniques include hydroprocessing, catalytic cracking of pyrolysis vapors, and so on. (Zhang et al., 2007).

In hydroprocessing, oxygen is removed as H_2O and CO_2 when bio-oil is hydrogenated in the presence of a catalyst (e.g., Co-Mo, Ni-Mo supported on alumina). The process is typically carried out in two steps (Holmgren et al., 2008). The first step substantially reduces the oxygen content and total acid number. The deoxygenated bio-oil is then further hydroprocessed to produce a fuel that is usable in automobiles. Properties of hydroprocessed bio-oil are compared with gasoline in Table 9.6.

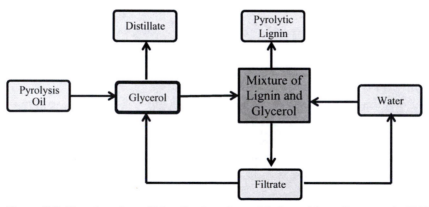

Figure 9.7. Fractionation of bio-oil using glycerol (adapted from Deng et al., 2009).

In catalytic cracking, oxygenated molecules in bio-oil are catalytically decomposed to hydrocarbons with the removal of oxygen as H_2O, CO, or CO_2. Examples of catalyst include ZnO, Al-MCM-41, and Cu/Al-MCM-41. Catalytic cracking increases the yield of acetic acid, furfural, and furans, and completely eliminates levoglucosan and large phenol molecules. Although catalytic cracking is a cheaper route for deoxygenation, it has a problem of high coke formation (8–25 wt%). The search for an efficient catalyst with high conversion and low coking is still on.

9.7 Summary

Depending on heat-transfer rate, the pyrolysis processes can be divided into two subclasses: slow pyrolysis (conventional pyrolysis or carbonization) and fast pyrolysis, out of which fast pyrolysis yields a high amount of liquid bio-oil. Available reactor designs include bubbling fluidized bed, circulating fluidized bed, vacuum pyrolysis, ablative fast pyrolysis, and auger reactor. Bio-oil is a viscous, corrosive, and unstable mixture of a large number of oxygenated molecules, depending on the pyrolysis process and biomass feedstock. Due to the high oxygen content, the heating value is less than half that of petroleum liquid. Bio-oil needs to be upgraded before use as liquid fuel. Various methods available include solvent fractionation, hydroprocessing, and catalytic cracking. Pyrolysis presents an attractive option to convert solid biomass into liquid bio-oil, which is easier to transport, store, and upgrade.

10 Biocrude from Biomass Hydrothermal Liquefaction

10.1 What Is Biocrude?

In search of renewable fuels, as early as the mid-twentieth century, researchers started to convert biomass into petroleum-like liquids. For example, Berl (1944) treated biomass using alkaline water at 230°C to produce a viscous liquid that contained 60% carbon and 75% heating value of the starting material. The liquid, termed biocrude, contains 10–20 wt% oxygen and 30–36 MJ/kg heating value as opposed to <1 wt% and 42–46 MJ/kg for petroleum (Aitani, 2004). The high oxygen content imparts lower energy content, poor thermal stability, lower volatility, higher corrosivity, and tendency to polymerize over time (Peterson et al., 2008). Hence, biocrude needs to be deoxygenated to make it compatible with conventional petroleum. Because biocrude contains less oxygen and more heating value than biomass, the hydrothermal liquefaction process is also referred to as hydrothermal upgrading. Compared with bio-oil from fast pyrolysis, biocrude produced from hydrothermal liquefaction has higher energy value and lower moisture content but requires longer residence time and higher capital costs. Typical hydrothermal liquefaction conditions range from 280 to 380°C, 70 to 20 Mpa, with liquid water present and reaction occurring for 10–60 minutes.

10.2 Hydrothermal Medium

Generally, hydrothermal medium refers to water that has been heated and compressed simultaneously. A phase diagram of pure fluid is shown in Figure 10.1; in the case of water, the critical point is 374°C and 221 bar (Note: a mixture may have a different critical point depending on the components and concentrations). As water is heated, it continues to acquire an interesting set of properties. For example, density decreases, dielectric constant decreases, and ionic product first increases and then decreases (Figure 10.2). Recently, subcritical and supercritical water have drawn significant attention to carry out chemical reactions due to the environmentally benign nature of the medium and its tunable physical properties. In addition, due to the increased solubility of gases and organics, many of the reactions can be conducted

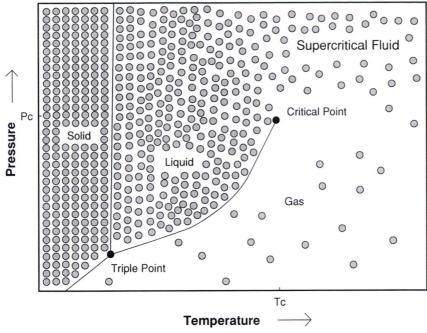

Figure 10.1. Schematic of molecular orientations in various phases of a pure substance.

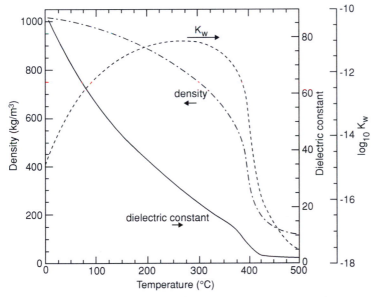

Figure 10.2. Density (Wagner and Pruss, 2002), static dielectric constant (Archer and Wang, 1990), and ion dissociation constant (K_w) (Bandura and Lvov, 2006) of water at 300 bar as a function of temperature (Peterson et al., 2008; reproduced by permission of The Royal Society of Chemistry).

in a single fluid phase that would otherwise occur in a multiphase system under conventional conditions. The advantages of a single-phase reaction medium are that higher concentrations of reactants can often be attained and there is no interphase mass-transport processes to hinder reaction rates (Phillip, 1999; Muthukumaran and Gupta, 2000; Kruse and Dinjus, 2007a, 2007b). In addition, the nonpolar products simply precipitate out upon cooling, providing a facile separation. Key advantages of these unique physicochemical properties are discussed below.

The dielectric constant drops drastically as water is heated and starts behaving more like a nonpolar solvent at near- or supercritical conditions.

10.2.1 Dielectric Constant

The dielectric constant can play a key role in influencing biomass reactions, as described by the Kirkwood relation. During a reaction, the transition state may be of lower or higher polarity than the initial state. A high dielectric medium lowers the activation energy of a reaction for a transition state of higher polarity than the initial state. As a consequence, many reactions have a high activation volume. By variation of the relative dielectric constant with temperature and pressure, reaction rates can be controlled (Yamaguchi, 1998; Phillip, 1999). The dielectric constant of hydrothermal medium can be continuously varied with temperature from 80 at ambient to about 5 at the critical point (Figure 10.2). An intermediate dielectric constant resembling organic solvent can also be achieved. For example, at 130°C (density 0.9 g/cm^3), the dielectric constant is around 50, which is near that of formic acid. At 200°C (density 0.8 g/cm^3), the dielectric constant of 25 is similar to that of ethanol.

10.2.2 Ion Product

Water undergoes self-ionization to produce hydronium (H_3O^+) and hydroxide ions (OH^-) as

$$2H_2O \leftrightarrow H_3O^+ + OH^-$$

This ionization is evident from the fact that, even in the purest form of water, slight electrical conductivity can be detected using a sensitive instrument. The ionization constant (K_w) is calculated from the concentrations as

$$[H_3O^+][OH^-]/[H_2O][H_2O]$$

and has a value of 1.0×10^{-14} at ambient condition. The value increases by 1,000-fold when liquid water is heated to near critical (200–300°C) conditions. Hence, near critical water provides an excellent acid/base environment, which can catalyze a number of biomass-related reactions, without the need for adding external reagents. However, as one exceeds the critical point, K_w decreases dramatically. For example, K_w is about 9 orders of magnitude lower at 600°C (and 25 MPa) than at ambient condition. Hence, supercritical water in this high-temperature, low-density region is more suitable for free radical reactions and a poor medium for ionic reactions. Acid- or base-catalyzed reactions in hydrothermal medium show a characteristic non-Arrhenius kinetic behavior near the critical point of water. Below the critical temperature of water, the reaction rates usually increase with temperature until the critical temperature is reached. At the critical point, reaction rates can decrease or increase drastically depending on the chemistry and properties (Masaru, Takafumi, and Hiroshi, 2004). This variation offers the possibility of using pressure and temperature to tune the properties of the reaction medium to optimal values for a given chemical transformation (Phillip, 1999).

10.2.3 Solubility of Organics

Although ambient water solubilizes polar compounds, near-critical water (200–300°C) dissolves both nonpolar organic molecules and inorganic salts and is comparable to that of the popular organic solvent acetone, mainly due to a reduced dielectric constant in the 20–30 range (Ragauskas et al., 2006), from that of 80 for ambient water, while keeping density high enough (0.7 to 0.8 g/cm^3). As the solubility of organics in water increases with the increase in temperature, the behavior is different with respect to inorganic salts. For example, the solubility of salts reaches a maximum at 300–400°C; after that, the solubility drops very rapidly with an increase in temperature (e.g., the solubility of NaCl water is 40 wt% at 300°C and 100 ppm at 450°C, 25.3 MPa). At supercritical temperatures, water practically has no dissolving power for salts due to a very low (near 2) dielectric constant. In addition, due to the dramatic drop in the dissociation constant, dissolved salts behave like weak electrolytes (Yesodharan, 2002).

10.2.4 Diffusivity and Viscosity

Additional useful properties of hydrothermal medium are the high molecular diffusion and low viscosity. Both give rise to efficient heat transfer and mass transfer for solid/liquid biomass liquefaction reactions. For example, at 15 MPa, viscosity of water at 25°C is 0.89 cP, at 200°C is 0.14 cP, and at 300°C is 0.09 cP. Diffusivity of molecules is not as high as in the gas phase but is much higher than in ambient water. Both of these properties can be used to reduce heating and mixing needs in the hydrothermal reactors. A combination of high diffusivity, low viscosity, and high miscibility can accelerate chemical reactions and improve reaction efficiency.

10.3 Liquefaction Process

Hydrothermal liquefaction of biomass using subcritical water has some potential advantages over other processes. Because water serves as both reaction medium and reactant, the process can use wet biomass without the need for energy-intensive drying. Moreover, biomass residue generated from many varieties of operations can

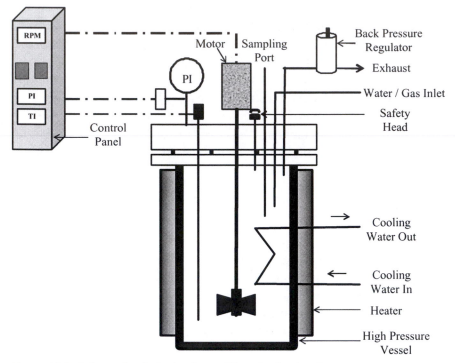

Figure 10.3. Schematic of a hydrothermal liquefaction batch reactor.

be directly used. The liquefaction has been tested in both batch and continuous processes.

10.3.1 Batch Process

Biomass is mixed with water and pressurized or pumped to a high pressure (typically 70–200 bar), and then the temperature is raised to 280–380°C. Catalysts can be added along with feed water to enhance the liquefaction. Reaction is carried out for 10–60 minutes, and then the product is cooled and separated. Feeding biomass slurries at high pressure is challenging, particularly at laboratory scale. Hence, a large number of experimental studies by researchers have been carried out using batch reactors. A hydrothermal liquefaction reactor is shown in Figure 10.3. The reactor consists of a high-temperature, high-pressure stirred-tank reactor equipped with temperature, pressure, and stirring controllers.

The biomass, water, and catalyst (if needed) are charged into the high-pressure vessel (reactor). As reactor is heated, pressure builds due to the expansion of water. If needed, pressure can be further increased by pumping in a gas (e.g., N_2, Ar, CO_2) or additional water. A magnetically coupled agitator is used for stirring the reaction mixture, if needed. After completion of the experiment, the reactor is cooled down to ambient temperature using cooling water connected to the reactor. The product mixture consists of solid, liquid, and gas fractions. Gas and liquid products are collected from the top and bottom ports, respectively, and the solid product

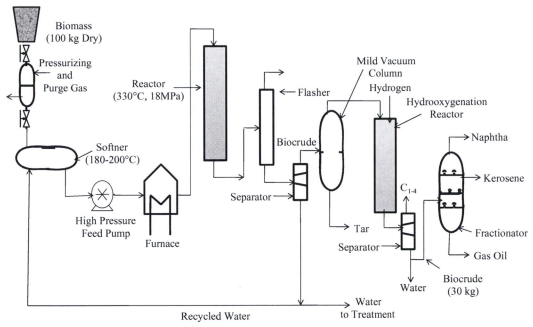

Figure 10.4. Continuous hydrothermal liquefaction process (adapted from Goudriaan and Peferoen, 1990; Feng et al., 2004a).

is collected by opening the reactor. Then, the biocrude production cycle is repeated. The batch reactor may suffice for the mobile biocrude units, where the unit is taken from one biomass site to another, and biocrude is transported to the central plant. However, for the large-scale, in-plant operations, a continuous unit should be used to reduce the capital cost and have heat integration.

10.3.2 Continuous Process

In the 1980s, a hydrothermal upgrading process was developed by Shell, for which typical conditions are shown in Figure 10.4 (Goudriaan and Peferoen, 1990). The biomass is first fed into a softener vessel via a lock hopper mechanism. In this vessel, the biomass is digested for 10–15 minutes in water at 180°C and is converted into a paste form so that it can be pumped to a high pressure. Using a slurry pump, biomass paste is pumped in the tubular reactor maintained at 180 bar and 330°C, where biomass spends about 7 minutes. Some of the heating in the reactor comes from the exothermic reaction itself. The reaction products are taken to a flasher to remove gases and then to a separator to remove biocrude from the excess water. A part of the aqueous stream is recycled back to the biomass softener. A mild vacuum distillation is used to remove tars from the biocrude. The biocrude is then sent to a hydrodeoxygenation reactor, where the remaining oxygen is removed as water. The hydrocarbon products are fractionated into naphtha, jet fuel, and diesel. Tar is used to provide process heating needs and hydrogen. In a conceptual process scheme, it was shown that each ton (dry basis) of biomass can produce 300 kg (or 95 gallons)

Table 10.1. *Product from hydrothermal liquefaction of biomass at 350°C, 180 bar, and 6 minutes residence time[a] (Goudriaan and Peferoen, 1990)*

	Mass ratio of water-to-biomass (dry basis)	
	7.33	3.0
Products (wt% on dry-ash-free feed)		
Biocrude	42.5	48.6
Gas	18.0	32.8
Water-soluble organics	}39.5	7.8
Water		10.8
Biocrude composition (wt%)		
C	75	81.9
H	6	8.1
O	19	10.0
Gas composition (wt%)		
CO_2	93	94
CO	6	2
CH_4	0.6	4
H_2	0.3	0.2

[a] Biomass contains 29–30 wt% lignin and 68 wt% cellulose + hemicelluloses

of liquid fuel. The typical composition of the products obtained from hydrothermal liquefaction is shown in Table 10.1.

10.3.3 Pumping Biomass with Biocrude

For ease of pumping, a high water-to-biomass ratio is desired. High water content in the liquefaction reactor causes an increase in the water-soluble organics and an associated decrease in biocrude yield. It is desirable to maintain a ratio of 3 (water-to-dry biomass) or lower. To solve this challenge, White and Wolf (1987, 1995) prepared biomass slurry of as high as 60 wt% using the recycled heavy portion (tar) of biocrude. In their operation, dry wood power was mixed with tar to produce slurry. The slurry was pumped using an extruder-feeder device similar to what is used in the polymer industry for pumping viscous liquids. The device consists of a helical screw, conveying the slurry through a barrel. The extruders can covey solids as long as the friction of the solids on the screw surface is not excessively greater than the friction of solids on the barrel surface. The pressure (about 200 bar) is maintained by using a valve at the end of the reactor.

In the work of White and Wolf (1987, 1995), wood powder/heavy biocrude slurry is pumped and mixed with water and a sodium carbonate catalyst. The reactor was maintained at 208 bar, 350°C, with a residence time of 80 minutes. The product is depressurized to yield biocrude having a near-ambient melting temperature and oxygen and dissolved water contents of 12–14 wt% and 3–5 wt%, respectively. Over the past decade, good-quality extruder for the manufacture of wood/polymer

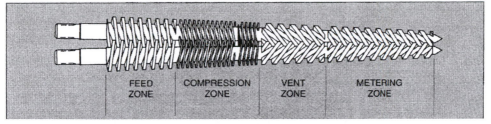

Figure 10.5. Conical twin screw in extruder for biomass (adapted from Maine, 2006).

composites has been developed. These use conical counter-rotating twin screws (Figure 10.5) and have shown to provide good performance. Additional developments have led to the development of high-pressure biomass feeders (Osato et al., 2004). The latest development can be used to pump biomass slurries into the reactor.

10.4 Liquefaction Mechanism

Liquefaction of biomass in subcritical water proceeds through a series of structural and chemical transformations (Chornet and Overend, 1985; Kumar and Gupta, 2009) that involve the following:

- Solvolysis of biomass resulting in micellar-like structure;
- Depolymerization of cellulose, hemicelluloses, and lignin; and
- Chemical and thermal decomposition of monomers to smaller molecules.

Although the product composition changes with the liquefaction conditions, an approximate overall reaction can be written as

$$\text{biomass (100 kg dry-basis)} \leftrightarrow \text{biocrude (48.6 kg)} + \text{gas (32.8 kg)}$$
$$+ \text{water-soluble organic (7.8 kg)} + \text{water (10.8 kg)} + 80 \text{ MJ}$$

with typical composition of gas as 94% CO_2, 2% CO, 4% CH_4, and 0.2% H_2 on mass basis. The overall hydrothermal liquefaction reaction is exothermic with heat of reaction of about -800 kJ/kg biomass (dry basis) (Goudriaan and Peferoen, 1990).

In hydrothermal liquefaction of biomass, water acts both as a solvent and reactant simultaneously (Yu, Lou, and Wu, 2008). An important reaction in the hydrothermal medium is the hydrolysis reaction, in which water reacts with biomass components to cleave bonds to open up the polymer structure. The hydrolysis is acid-catalyzed; hence, the high dissociation constant of water in the 200–300°C range is useful, which gives a high concentration of hydrogen ions. Additionally, CO_2, a product from biomass reactions, dissolves in water and increases the availability of protons. At higher temperatures, the hydrolysis reactions are often accompanied by pyrolysis reactions (Brunner, 2009).

In addition to the acid-catalyzed pathway, the hydrolysis proceeds with a direct nucleophilic attack mechanism, a hydroxide ion-catalyzed mechanism, and a radical mechanism. The source of hydrogen and hydroxide ions is the self-dissociation of water molecules, which makes water in hydrothermal medium a Brønsted base-acid and acts as an effective catalyst (Jin, Zhou, and Takehiko, 2005). A detailed description of reaction steps in the decomposition of biomass in hydrothermal medium is challenging, because numerous components that arise from cellulose, hemicelluloses, and lignin interact with each other, leading to a very complex set of chemical reactions (Minowa, Zhen, and Ogi, 1998, 1999). Additionally, the process is heterogeneous proceeding inside and, in particular, on the surface of biomass particles.

Hemicelluloses are polysaccharides of five-carbon sugars, such as xylose, or six-carbon sugars other than glucose. They are usually branched and have a much lower degree of polymerization. Branches in the chains do not allow for the formation of tightly packed fibrils. Hemicelluloses are not crystalline and are easily hydrolyzable to their respective monomers. In fact, about 95% of hemicelluloses were extracted as monomeric sugars using water at 200–230°C in a span of just a few minutes (Mok and Antal, 1992; Kumar and Gupta, 2009). In hydrothermal medium, hemicelluloses are hydrolyzed to sugars, which subsequently degrade into furfural and other degradation compounds. Furfural (2-furaldehyde) is commercially produced from hemicellulose-derived xylose (Peterson et al., 2008).

Hydrothermal degradation of cellulose is a heterogeneous and pseudo-first-order reaction for which detailed chemistry and mechanism have been proposed (Bobleter, 1994). Cellulose reaction in hydrothermal and catalyst-free medium mainly proceeds via hydrolysis of glycosidic linkages. The long chain of cellulose starts breaking down to smaller molecular weight water-soluble compounds (oligomers) and further to glucose (monomer). Glucose is water-soluble and undergoes rapid degradation in hydrothermal medium at elevated temperatures (Kumar and Gupta, 2008). Glucose further epimerizes to fructose or decomposes to erythrose plus glycolaldehyde or glyceraldehydes plus dihydroxyacetone. Produced fructose further decomposes to erythrose plus glyceraldehydes or glyceraldehyde plus dihydroxyacetone. Glyceraldehyde converts to dihydroxyacetone, and both glyceraldehyde and dihydroxyacetone dehydrate to form pyruvaldehyde. Pyruvaldehyde, erythrose, and glycolaldehyde can further decompose to smaller organic species, mainly acids, aldehydes, and alcohols of 1–3 carbon atoms (Sasaki et al., 2002). At nearly 350°C at 25 MPa, cellulose hydrolysis is faster than glucose degradation rate. For short-residence time experiments, one may even see a build up of glucose. Studies with model compounds have shown that aromatics can also be formed from sugar compounds derived from cellulose hydrolysis (Nelson et al., 1984; Luijkx, Rantwijk, and Bekkum, 1993). A network of cellulose liquefaction reactions is shown in Figures 10.6.

Lignin is a complex and high-molecular-weight polymer of phenylpropane derivatives (p-coumaryl alcohol, coniferyl alcohol, and sinapyl alcohol). The density of hydrothermal medium is found to be a key parameter in lignin decomposition. In hydrothermal reaction medium, most of the hemicelluloses and part of the lignin are solubilized below 200°C. Lignin fragments, due to their high reactivity, again cross-link and recondense to form high-molecular-weight, water-insoluble products (Bobleter, 1994). In supercritical water, lignin can completely dissolve and undergo homogeneous hydrolysis and pyrolysis. Various pathways for lignin reactions are

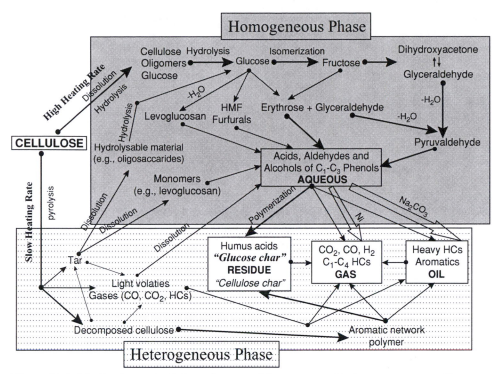

Figure 10.6. Mechanisms of cellulose decomposition promoted in homogeneous and heterogeneous environments in hydrothermal medium (Fang et al., 2004; reproduced with permission of the American Chemical Society).

shown in Figure 10.7 for all four phases (oil, aqueous, gas, and solid) present in the hydrothermal reactor (Fang et al., 2008).

10.4.1 Hydrothermal Treatment with Catalysts

Due to their low cost and high activity, alkali salts are the most used catalyst in hydrothermal liquefaction. Formate ions derived from the alkali carbonates react with the hydroxyl group of biomass to cause decarboxylation, which forms esters. Thus, alkali carbonates catalyze hydrolysis of biomass macromolecules into smaller fragments. The micellar-like, broken-down fragments produced are then degraded to smaller compounds by dehydration, dehydrogenation, deoxygenation, and decarboxylation. These compounds further rearrange through condensation, cyclization, and polymerization, leading to newer compounds (Demirbas, 2000a). However, with the choice of an alkali, there are some differences in activity. For example, Karagoez et al. (2005) examined the effect of various alkali salts on the yield of biocrude from the hydrothermal liquefaction of pine wood (Table 10.2). Based on the conversion and yield of liquid products, the catalytic activity ranks as: $K_2CO_3 >$ $KOH > Na_2CO_3 > NaOH$. For example, in noncatalytic liquefaction, the solid residue was about 42 wt%, whereas 0.94 M K_2CO_3 gave a more complete conversion

Table 10.2. *Effect of 0.94 M K_2CO_3 on hydrothermal liquefaction of wood at 280°C for 15 minutes (Karagoez et al., 2005)*

Catalyst	Biocrude (wt%)	Gas (wt%)	Solid residue (wt%)	Water-soluble organics (wt%)
None	8.6	9.7	41.7	40.0
NaOH	22.4	12.0	14.0	51.6
Na_2CO_3	23.0	11.5	11.5	54.0
KOH	28.7	11.8	8.6	50.9
K_2CO_3	33.7	11.0	4.0	51.3

with residue of only 4.0 wt%. In noncatalytic experiments, furan derivatives were observed, whereas the catalytic runs produced mainly phenolic compounds.

Additional catalysts, including various organic and inorganic acids (H_2SO_4, HCl, acetic acid), metal ions (Zn^{2+}, Ni^{2+}, Co^{2+}, and Cr^{3+}), salt and bases [$CaCO_3$, $Ca(OH)_2$, HCOONa, and HCOOK], and CO_2 have been tested with varying degrees of success (Miyazawa and Funazukuri, 2005). For example, Minowa et al. (1998, 1999) compared sodium carbonate with reduced-nickel catalyst in the 200–350°C temperature range, to conclude the role of alkali catalyst in inhibiting the formation of char from biocrude (stabilization of biocrude), resulting in more

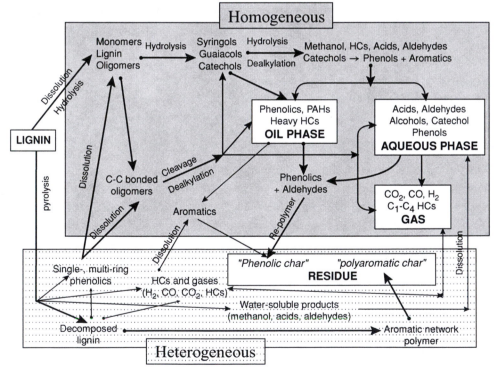

Figure 10.7. Reaction paths of lignin decomposition promoted in homogeneous and heterogeneous environments in hydrothermal medium (Fang et al., 2008; reproduced with permission of Elsevier).

biocrude production. On the other hand, the nickel catalyst catalyzed the steam-reforming reaction of aqueous products as intermediates and the methanation reaction (Minowa et al., 1998, 1999).

10.4.2 Hydrothermal Treatment with Reducing Gases

Reducing gases, such as CO and H_2, have been investigated in the early liquefaction experiments, in hopes of further reducing the oxygen content of biocrude (Appell, 1977; Davis, Figueroa, Schaleger, 1982). But recently, experiments have shown that the effect is very minor (He et al., 2001). On the other hand, intentional addition of CO_2 to the reactor causes a noticeable increase in the oxygen content of biocrude.

10.5 Properties of Biocrude

Biocrude is solid at room temperature and is typically liquid at temperatures above 80°C. A large portion is chloroform (or acetone)-soluble, with a small amount of insoluble material of high molecular weight. Oxygen content is typically 10–20 wt% and is furanic or aromatic in nature. The heating value is in the range 30–36 MJ/kg. In hydrothermal liquefaction, a small content of organics and heating value remains in the aqueous phase, which should be used or recycled for full recovery.

Biocrude contains a mixture of a large number of molecules with molecular weight as large as 2000. These components can be classified into organic acids, alcohols, aldehydes, ketones, aromatic hydrocarbons, aliphatic hydrocarbons, furans, and phenols. Some of the prominent compounds in the biocrude from corn stover are phenol, guaiacol, 4-ethyl-phenol, 2-methoxy-4-methyl-phenol, 4-ethyl-guaiacol, 2,6-dimethoxyphenol, 1,2,4-trimethoxybenzene, 5-tert-butylpyrogallol, 1, 10-propylidenebis-benzene, 1-(4-hydroxy-3,5-dimethoxyphenyl)-ethanone, acetic acid, 1-hydroxy-2-propanone, furfural, 3-methyl-2-cyclopenten-1-one, 2,5-hexanedione, and desaspidinol (Zhang, Keitz, and Valentas, 2008). Typically, the biocrude distillate contains 19–23% aromatic hydrocarbons, 11–35% phenols, 9–13% naphthenes, 5–11% aliphatic hydrocarbons, 7–14% aldehydes and ketons, and 3–10% alcohols, on mass basis (Adjaye, Sharma, and Bakhshi, 1992).

Viscosity of biocrude varies greatly with the liquefaction conditions. For biocrude from the batch liquefaction with 5:1 water-to-biomass ratio and use of 10 wt% (of biomass) sodium carbonate catalyst, the viscosity at 25°C is 313×10^6 cP, which increased to 323×10^6 cP upon storage for a month. The high and increasing viscosity indicates a poor flow characteristic and stability. The increase in the viscosity can be attributed to the continuing polymerization and oxidative coupling reactions in the biocrude upon storage. Although stability of biocrude is typically better than that of bio-oil, the viscosity of biocrude is much higher. For example, biocrude and bio-oil obtained by Elliott and Schiefelbein (1989) are compared in Table 10.3. Density of biocrude is near 1 g/cm^3 and remains essentially constant upon storage.

Table 10.3. *Properties of biocrude from hydrothermal liquefaction compared with bio-oil from fast pyrolysis (Peterson et al., 2008; Elliott and Schiefelbein, 1989)*

	Fast pyrolysis	Hydrothermal liquefaction
Higher heating value (MJ/kg)	22.6	35.7
Viscosity (cP)	59 at 40°C	15,000 at 61°C
Moisture content (wt%)	25	5
Elemental analysis (wt% dry basis)		
C	58	77
H	6	8
O	36	12

10.6 Refinement and Upgrading of Biocrude

The two key properties of biocrude that differ from the petroleum crude are high oxygen content and molecular weight (or viscosity). For compatibility with the petroleum liquids, biocrude needs to be upgraded by hydrogen treatment. The addition of hydrogen to the biocrude molecules should be performed not to saturate the aromatic rings, but to remove oxygen and hydrocrack molecules. Hence, the hydrodeoxygenation (HDO) process is used, which is somewhat different than the hydro-treating used in the petroleum industry to remove nitrogen and sulfur.

10.6.1 Hydrodeoxygenation

For HDO, hydrogen and biocrude are fed to a catalytic reactor maintained at 250–400°C and 10–18 MPa. Various catalysts have been successfully tested, including CoMo, NiMo, NiW, Ni, Co, Pd, and CuCrO, on a variety of supports, such as Al_2O_3, γ-Al_2O_3, SiO_2-Al_2O_3, and Y-zeolite/Al_2O_3. In a catalytic HDO of biocrude, oxygen content was reduced from 10 to <0.1 wt% with hydrogen consumption of 4 wt% on intake and C_1–C_4 gas production of only 2.4 wt% (Goudriaan and Peferoen, 1990). The product included liquids similar to diesel, jet fuel, and a small portion of naphtha, with a cetane number of >50.

Out of various catalyst choices, a promising candidate is the sulfided form of CoMo catalyst due to its high activity. However, the product fuel may become contaminated with sulfur, and the catalyst may get deactivated by coke deposition, and, mostly notably, may get poisoned by trace amounts of water present (Laurent and Delmon, 1994; Elliott, 2007). Recent development in the reductive upgrading of biocrude offers an attractive alternative route, which allows upgrading in the aqueous medium (Diaz et al., 2007; Zhao et al., 2009). In fact, the separation from aqueous media would also be facile due to the nonmiscibility of hydrocarbons upon cooling to ambient temperature (Figure 10.8).

In aqueous HDO (Zhao et al., 2009), biocrude is added to acidic aqueous medium along with Pd/C catalyst, and hydrogen is fed to the reactor. Typically, the reaction is carried out at 200–250°C for 30 minutes at 5 MPa. The oxygen is removed as H_2O, CH_3OH, and other oxygenates. A high conversion (about 100%)

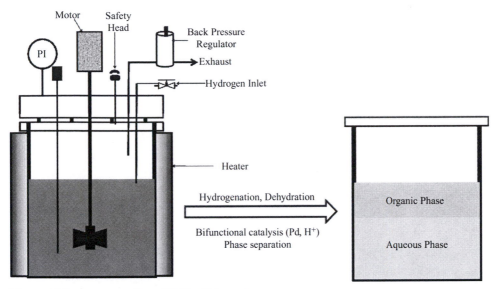

Figure 10.8. Aqueous-phase HDO of biocrude.

and high selectivity toward hydrocarbons are achieved. To illustrate, yields from various biocrude phenolic components are shown in Table 10.4.

10.7 Critical Issues

The hydrothermal liquefaction requires the use of high pressures, which necessitates the design of special reactors and separators, requiring a high capital investment. Despite several key benefits, the widespread commercial use of hydrothermal technology has been hindered due to the high capital cost and a few critical issues that need to be resolved (Peterson et al., 2008).

10.7.1 Heat Integration

Hydrothermal liquefaction operates at a high temperature and has a high heating requirement due to the high heat capacity of the dense medium. Hence, it is important to recover heat from the hot effluent streams to preheat the cold incoming streams. Heat integration is needed that optimizes the external heating requirements. In addition, a drop in heat exchange rate over time due to fouling needs to be accounted for. A heat exchange network (Figure 10.9) has been proposed by Feng, van der Kooi, and de Swaan Arons (2004b), which allows for an efficient operation of the combined reaction and separation operations.

10.7.2 Biomass Feeding and Solids Handling

The majority of the hydrothermal liquefactions tests have been performed with "clean" biomass feedstock, such as ground wood or model compounds. However, commercial feedstocks are expected to be mixed waste streams, including sewage,

Table 10.4. *Selectivities (%) in aqueous-phase HDO of biocrude-related phenolic compounds at 250° C using Pd/C catalysts with acids (Zhao et al., 2009)*

				Methanol and ketones
OH, OCH₃, CH₂CH₂CH₃	66	3.0	6.7	24.2
OH, OCH₃, CH₂CH=CH₂	65	3.5	5.0	25.4
OH, OCH₃, CH₂COCH₃	71	3.1	5.2	20.2
OH, H₃CO, OCH₃, CH₂CH=CH₂	58	3.6	5.4	33

sludge, manure, and dirt. The handling of inorganic impurities can become crucial as these can create problems due to precipitation, scaling, clogging, and plugging catalyst pores. Otherwise, the separation and purification of feedstock can significantly contribute to the cost.

The water-to-biomass feed ratio to the reactor is another concern. It is easier to pump low-biomass loadings. But, at low concentrations, the liquefaction process is expensive due to the high capital cost of heat exchangers, pumping expenses, and heat losses. Hence, biomass loading in excess of 15 wt% (dry biomass basis) is required to achieve practical economies. Unfortunately, feeding at high loading in a reactor operating at 20 MPa is challenging. Some of the emerging concepts

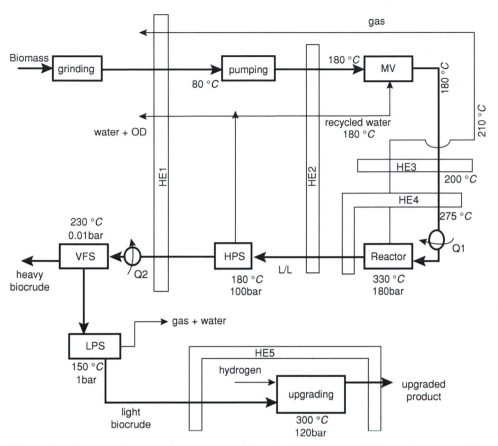

Figure 10.9. Proposed heat exchange network for the HTU process. MV, mixing vessel; HE, heat exchanger; HPS, high-pressure separator; LPS, low-pressure separator; VFS, vacuum-flash separator (Feng et al., 2004b; reproduced with permission of Elsevier).

to solve this challenge are using starch with cement pumps, prehydrolyzed feed of paste consistency, pumping water against a piston containing biomass slurry, and the latest use of a conical counter-rotating, twin-screw feeder.

10.7.3 Recovery of Inorganics and Catalysts

Biomass is composed of C, H, O, N, P, S, K, and Na, along with a number of other elements in small concentrations (i.e., micronutrients). Carbon and hydrogen elements are used for producing the fuel. Oxygen is discarded as H_2O and CO_2. The remaining elements, especially N, P, and K, have a high commercial value if they can be recovered in biologically active (e.g., NH_4^+, NO_3^-, K^+, PO_4^{3-}, etc.) forms for use as fertilizers. If these ions are precipitated out and recovered from the hydrothermal liquefaction process, then produced NPK fertilizer will add to the sustainability of the biomass production.

Solid catalysts that remain in the reactor as the biocrude flows past them are prone to inactivation by fouling, coking, precipitation of other inorganics, and fixing of sulfates. Maintenance of the catalyst activity and the recovery of used catalysts need to be properly designed to ensure continuous and long-term operation of the plant. Soluble catalysts (e.g., KOH, Na_2CO_3, H_3PO_4, H_2SO_4) offer the advantage of not suffering from coking and inactivation problems. However, these catalysts must be recovered and reused, if expensive, at the end of the process in order to achieve an economic liquefaction.

10.7.4 Reactor Wall Effects

During hydrothermal liquefaction, significant amounts of organic acids are produced, which can cause precipitation of inorganic salts and catalysts onto the reactor wall. These can bring serious corrosion to the reactor. Hence, often the reactors are constructed of high-nickel alloys. Because nickel catalyzes some of the hydrothermal reactions, the reactor wall itself can act as a catalyst, and its effect may be difficult to separate from the intentionally added catalyst. Such effects may lead to a scale-up issue if not properly understood.

10.8 Summary

Hydrothermal treatment can be used to liquefy biomass and increase energy density to produce biocrude. Typical hydrothermal liquefaction conditions range from 280 to 380°C, 7 to 20 MPa with liquid water present, and the reaction occurs for 10–60 minutes. Both batch and continuous reactor schemes are available, with some challenges in continuous feeding of biomass. Recent advancements (e.g., twin-screw feeding) are likely to address some of the challenge. Compared with petroleum, the biocrude has its own characteristics, such as higher oxygen content, consisting of various types of molecules of wide-ranging molecular weight, and being in solid state at ambient temperature. Biocrude has an oxygen content of 10–20 wt% and heating value of about 35 MJ/kg, which can be further improved by HDO to produce liquids similar to diesel and jet fuel. The process has a high heating requirement; hence, proper energy integration is needed for the commercial plants.

11 Solar and Wind Energy for Biofuel Production

11.1 Process Energy Needs for Biofuel Production

Biofuel production has a variety of process heating needs depending on the conversion route adopted (Figure 11.1). In overall, the biomass thermochemical conversion processes are endothermic, and the heat required can be supplied by concentrated solar energy in such a way that the energy evolved from the fuel produced ideally represents the sum of energy stored during the photosynthesis and the direct thermal collection (Lede, 1998). Some of the reaction steps are exothermic in nature, which can provide part of the energy needed for the other steps in the plant. Nonetheless, if extra energy is supplied from solar and wind resources, then more of the biomass carbon can be converted into the liquid fuels.

Drying of biomass can put a heavy heat load on the process. For example, the external heating needs are about 3000–4000 kJ/kg of water removed (Lede, 1998).

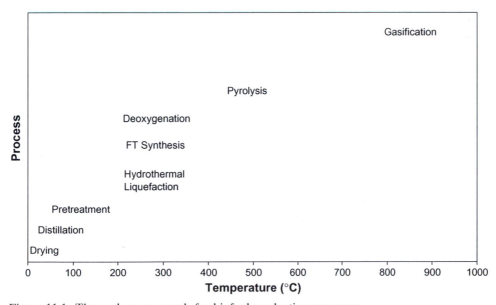

Figure 11.1. Thermal energy needs for biofuel production processes.

175

Table 11.1. *Electric energy requirement for grinding of biomass (with 12 wt% moisture content), for hammer mill with various size screen openings (Mani et al., 2004)*

Biomass	Specific electrical energy consumption (kWh/ton)		
	3.2-mm screen	1.6-mm screen	0.8-mm screen
Wheat straw (7.7-mm size)	24.7	43.6	45.3
Barley straw (20.5-mm size)	n/a	27.1	99.5
Corn stover (12.5-mm size)	11.0	19.8	34.3
Switchgrass (7.2-mm size)	27.6	58.5	56.6

Hence, for a 1-kg water + 1-kg dry biomass feedstock, it will cost about 3000–4000 kJ in drying to get dry biomass with a heating value of 15,000 kJ; therefore, the drying needs are about 20–27% of the energy content of biomass.

The electrical (or mechanical) energy needs of the process are for conveying, grinding, and pumping. Energy needed for biomass grinding depends on the initial biomass size, final particle size, moisture content, material properties, scale of operation, and machine variables (Mani et al., 2004). Some of these values for four biomasses are shown in Table 11.1. For example, for about 7 mm original biomass size, using 0.8 mm screen size, the specific energy consumption by hammer mill for switchgrass is 56.6 kWh/ton and that for barley straw is 99.5 kWh/ton, which correspond to 203.8 and 358.2 kJ/kg electric energy, respectively. For reference, the heating value of biomass is about 15,000–17,000 kJ/kg, and typical conversion of thermal to electric power is only 25% efficient. Hence, 203.8 kJ of electric power is equivalent to about 815.2 kJ of heat, representing about 5% of the energy content of switchgrass.

In the following sections, various aspects of wind (for process electricity and hydrogen needs) and solar (for process heating needs) energies are discussed.

11.2 Wind Energy

Wind energy has been used for many centuries to pump water and mill grains. As opposed to the solar heat, it is a form of mechanical energy that can be efficiently converted to electrical energy. Also, wind systems can produce electricity 24 hours every day (both during night and day time, as long as the wind is blowing), unlike solar systems that cannot provide heat or electricity during the night. That is why wind power is promising in coastal and windy areas. But, in other geographical areas, wind energy has environmental limitations (e.g., dead birds, noise). Overall, wind power has experienced a dramatic growth in recent years. For example, worldwide installed capacity increased to 120,798 MW in 2008, compared with only 39,431 MW in 2003 (GWEC, 2009). The majority of the installations are in the United States, Germany, Spain, China, and India, with 20.8%, 19.8%, 13.9%, 10.1%, and 8.0% of the global capacity, respectively.

A wind turbine consists of a tower, which carries the nacelle (housing), and the turbine rotor consisting of rotor blades and hub. The modern turbines have three

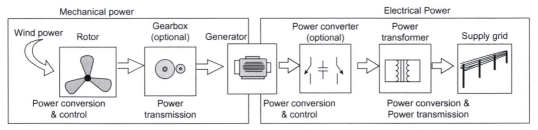

Figure 11.2. Main components of a wind turbine system (Chen and Blaabjerg, 2009; reproduced with permission from Elsevier).

rotor blades placed upwind of the tower and the nacelle. Anemometers and wind wane to measure wind speed and direction, and aviation lights are placed outside the nacelle (Chen and Blaabjerg, 2009). The key components (i.e., gearbox, mechanical brake, electrical generator, control system, etc.) are contained in the nacelle. The main components of a modern wind turbine system are illustrated in Figure 11.2, including the turbine rotor, gear box, generator, transformer, and possible power electronics (Chen and Blaabjerg, 2009).

With the recent advances in wind turbine technology, the cost of wind electricity has dropped significantly over the last 20 years. For example, in the early 1980s, the wind electricity cost was about $0.30/kWh. But now, some of the excellent sites with the latest wind power plants can produce electricity at a much lower cost. It is expected that the cost will continue to decrease as larger-scale plants are built and the advanced technology is adopted (AWEA, 2005). The major factors that influence the cost are the size of the wind farm, wind speed at the site, cost of installing the turbines, and turbine efficiency. Recent estimates suggest that the power generation cost of an onshore wind farm is between $0.06 and $0.12/kWh and that of an offshore wind farm is between $0.08 and $0.16/kWh, with the number of full hours and the level of capital cost being the most influencing factors (Blanco, 2009).

In addition to electric power, the biofuel plant also needs hydrogen for hydrodeoxygenation of bio-oil and biocrude, and for conditioning of syngas. Hydrogen can be derived from using some of the biofuel, but if external hydrogen is available, then biofuel yield will be high. Wind power can contribute toward both electricity and hydrogen needs of the plants. The extra power during peak hours is used to electrochemically split water into hydrogen and oxygen, as shown in Figure 11.3. In fact, electric energy conversion to hydrogen energy can help with the load-balancing problems that arise due to fluctuations in wind speed. Hydrogen is stored onsite and used in the biofuel plant when needed (Gupta, 2008).

Not all biomass production locations have enough wind speed. But there are a large number of farms that can install wind energy systems. A typical wind turbine removes less than an acre of land from production. And for each wind turbine they host on their land, farmers may receive income in the range of $2,000–$10,000 per year in lease or royalty payment (Bolinger and Wiser, 2006). The payment may further improve if the wind electricity is locally used for biofuel processing, due to more

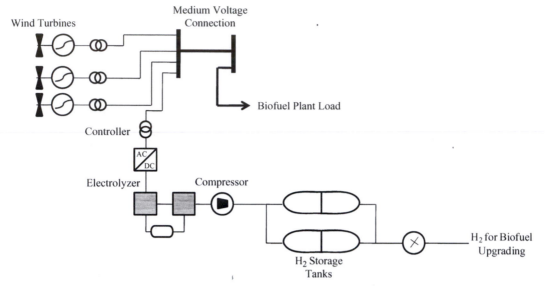

Figure 11.3. Schematic of power and hydrogen production from wind energy.

complete use of the energy resources and load balancing. Hence, with biomass production, farmers may be able to produce two forms of renewable energies: biomass and wind power.

11.3 Solar Energy

Sunlight is one of the most abundant energy resources on the surface of the earth. In fact, biomass is a form of solar energy that has been stored as carbohydrates. Hence, the places that have a high production of biomass also have a high solar radiation, provided there is enough rain or water supply. However, both energy sources are in diffused form (not concentrated as in coal mine or petroleum well). In most locations, solar energy is available near biomass; therefore, the coupling of the two energy sources can be achieved with ease. In some of the past demonstrations, solar energy was used directly to provide heat to the process; whereas, in other cases, intermediate heat exchange media have been used to carry the energy to the process.

11.3.1 Solar Collectors

The key part of the heat supply from solar energy is the solar collectors, which convert solar radiation into heat using reflector/collector and receiver/reactor. The concentration of solar radiation is done by sun-tracking mirrors called collectors (or heliostats), which focus the radiation on a solar receiver/reactor. Over the last 30 years, a large number of solar concentrators have been developed (Figure 11.4); out of these, four designs have come to prominence: (1) parabolic trough, (2) central power tower, (3) parabolic dish, and (4) double concentration (Kodama, 2003).

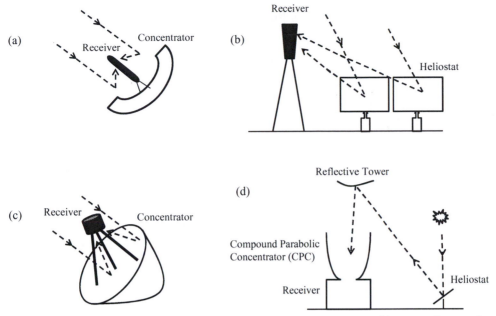

Figure 11.4. Main concepts of large-scale solar concentrating systems: (a) parabolic trough, (b) central power tower, (c) parabolic dish, and (d) double concentration (adapted from Kodama, 2003).

In the parabolic trough system, sunlight is focused along the focal line of the linear parabolic concentrator (Figure 11.4a). Solar energy is absorbed by a working fluid (typically steam or heat-transfer oil), which can be piped to the biofuel plant. In the parabolic trough, solar energy is concentrated by 30- to 100-fold, and the temperature of the working fluid can be achieved in the range of 225–425°C.

In the central power tower system, a field of tracking mirrors reflects the solar radiation onto a receiver, which is mounted atop a centrally located tower (Figure 11.4b). The energy in the receiver is absorbed by a working fluid (typically air or molten salt), which can reach temperatures up to 1,300°C. Depending on the number of two-axis tracking mirrors used, the solar concentration ratios of 300–1500 can be achieved.

In the parabolic dish system, a parabolic dish is used to concentrate solar radiation to a thermal received (Figure 11.4c). Depending on the size of the dish, concentration ratios of 1000–5000 can be achieved, giving a working fluid temperature of as high as 1,500°C.

In the double-concentration system, a heliostat field focuses on a reflective tower, which sends radiation to the ground and is received by a secondary concentrator (Figure 11.4d). Due to the double effect, a concentration of 5,000- to 10,000-fold occurs. Temperatures are received in excess of 1,200°C.

Thus, by using the last three designs (Figure 11.4b–d), high-flux radiation of 1,000–10,000 kW/m^2 can be achieved in the sunbelt region. The use of solar heat in biofuel production enables solar energy storage and transportation to the population centers (Kodama, 2003).

11.4 Direct Use of Solar Radiation

There have been a number of tests where solar radiation has been directly on biomass (Antal et al., 1981; Lede, 1998). These efforts have been for gasification and some for pyrolysis process.

11.4.1 Gasification

In usual gasifiers, the heat is supplied by several oxidation reactions by partially combusting biomass with air. Up to 30% of the biomass is used in supplying this heat. In addition, the produced syngas is diluted by nitrogen present in the air and the combustion product carbon dioxide (CO_2). If pure oxygen is used instead of air, the input costs are high. There are several benefits if the oxygen is supplied by water and gasification heat is supplied by solar radiation, including (1) complete use of biomass carbon for liquid fuel production, (2) no need for external oxygen supply, and (3) high calorific value of the syngas, as it will have less CO_2 and no nitrogen (N_2). Most of the gasification processes using solar heat have used small particles in fluidized-bed or free-falling reactors with direct illumination by concentrated solar radiation via transparent windows (Murray and Fletcher, 1994). Some studies have considered the use of massive biomass (e.g., whole logs). For example, the cross-section of a cylindrical wood log can be set at the focal zone of a concentrator and continuously fed as the wood is consumed (Lédé et al., 1983). Additional similar experiments have been conducted with packed beds of biomass adjusted at the focal zone (Taylor, Berjoan, and Coutures, 1980). The transparent windows typically perform well as the deposited char and tar are gasified due to the high intensity of radiation. The syngas has a composition of 50% hydrogen and 40% carbon monoxide, with a small amount of light hydrocarbons.

11.4.2 Pyrolysis

In fast pyrolysis for bio-oil production, needed heat is provided by combustion of gas and/or char fraction. Additionally, an external biomass or fossil fuel can be used. An intermediate heat carrier transfers the heat to the wall or directly to the biomass particles. For a good bio-oil yield, a very-high-heat flux is needed, which necessitates the use of a high temperature gradient. Use of solar radiation can favorably satisfy the thermal needs of fast pyrolysis (Lédé et al., 1980). Both conditions of high temperature and fast heating rate can be met by concentrated solar radiation. The solar flux can be concentrated inside the reactor in a small focal zone, which brings a new benefit. The pyrolysis gases liberated from the focal area quickly travel out to the cooler zones, avoiding the secondary cracking reactions. However, in experimental runs, some difficulties have been observed as discussed below (Lédé, 1998).

1. Due to the poor mass transfer, bio-oil droplets form a smoke screen between the window and the biomass. This decreases the radiation flux reaching the biomass, which in turn leads to slow pyrolysis condition, resulting in the formation of

charcoal. Additionally, the product in the smoke droplets undergoes secondary reactions (Lédé and Pharabod, 1997; Gronli, 1996).

2. Some of the biomass components (e.g., cellulose) do not absorb radiation efficiently, due to their high reflectivity (Hopkins and Antal, 1984; Boutin, Ferrer, and Lédé, 1998).

3. In fluidized-bed reactors, the biomass particles may enter the focal zone several times, and hence they undergo the cycles of flash heating and cooling. Such treatment leads to secondary reactions and charcoal formation. Also, some particles directly heated by radiation cause screening effects for other particles that receive much less heating flux.

Some of the above disadvantages can be overcome by combining some of the conditions of concentrated solar energy and ablative pyrolysis. For example, radiation can be provided in short flashes followed by quench time. Also, mechanical devices can be used to remove primary pyrolysis products from the window surface as soon as they are formed.

11.4.3 Challenges with Use of Direct Solar Radiation

In addition to the scale-up and modeling problems, there are challenges associated with the use of direct solar radiation in biomass processing, as summarized below (Lédé, 1998):

1. The transparent window needs to operate for a long time in the harsh chemical environment with multiple phases. The produced tar, oil, char, and ash may deposit or condense on the window, reducing the transparency for incoming solar radiation. The available flux for biomass decreases and the window heats, which can break the window. Injecting steam on the window can perform continuous cleaning, but can also alter the fluid dynamics of the reactor.

2. Due to the presence of multiple phases in the reactor, gravity and placement (e.g., horizontal versus vertical) can have a significant role in the operation. This can be a special challenge when using solar radiation, which is in constant directional change during the whole day. For example, when using the parabolic dish concentrators, the reactor position changes continuously while tracking the sun. In the case of solar towers, conveying biomass may be difficult from the ground to the reaction zone.

3. Intermittency of the solar radiation (due to cloud cover, daylight hours) causes problem with process stability and control. For example, in fast pyrolysis occurring for fractions of seconds, it is very difficult to adjust to sudden variations in solar illumination.

11.5 Storage of Solar Thermal Energy

Indirect heating can solve many of the challenges associated with direct heating with solar radiation. Indirect heating can be done using a heat-transfer fluid that

Table 11.2. *Potential materials for storage of thermal energy (Demirbas, 2006a)*

Material	Melting temperature (°C)	Heat of fusion (kJ/kg)
High-density polyethylene	100	200.0
Naphthalene	118	89.8
$MgCl_2.6H_2O$	117	168.6
Trans-1,4-polybutadiene	145	144.0
$NaNO_3 + KNO_3$ (60:40 ratio)	220–290	
$NaNO_3$	307	172.0
KNO_3	333	266.0
KOH	380	149.7
$MgCl_2$	714	452.0
NaCl	800	492.0
Na_2CO_3	854	275.7
KF	857	452.0
K_2CO_3	897	235.8
Zinc	420	112.0
Aluminum	660	397.0

carries stored thermal energy to the biofuel reactor. Solar energy is only available during the daytime, and the amount of radiation depends on the cloud cover and the season; therefore, it is necessary to store the thermal energy so that it can be continuously supplied to the biofuel production process. Thermal energy can be stored by heating a medium and/or by changing the phase of the medium. Usually phase change (e.g., solid to liquid or liquid to vapor) can accommodate more energy then simply sensible heating (Gutherz and Schiller, 1991; Hawes, Feldman, and Banu, 1993). However, liquid-to-vapor phase-change materials are not practical due to the large volumes or high pressures required to store the vapors. Additionally, thermal energy can be stored using endothermic reactions that can reversibly discharge the heat when needed. The phase-change material should be selected so that heat is stored at a temperature suitable for its use. Some of the phase-change materials suitable for heat for biofuel processing (>100–1,000°C) are listed in Table 11.2.

11.6 Summary

Solar energy has the potential to provide many of the thermal needs of biofuel production plants. The use of direct concentrated radiation in biomass reactors has been tested, with some challenges. A variety of solar systems are available that can produce and store energy at the temperature levels needed in biomass processing. A working fluid can then transfer the heat to the biofuel process when needed. Wind energy can provide some of the electricity and hydrogen needs. With supplementing energy from solar and wind sources, a higher amount of carbon in biomass can be converted to liquid fuels.

12 Environmental Impacts of Biofuels

12.1 Biomass and the Natural Carbon Cycle

In the natural carbon cycle, carbon is exchanged among the biosphere, pedosphere, geosphere, hydrosphere, and atmosphere of the earth. During photosynthesis, atmospheric carbon dioxide (CO_2) is converted into plant biomass. When carbon in the biomass is combusted or used for energy and materials for industry and agriculture, the biomass carbon is released back in different ways, but mostly as CO_2 to the atmosphere. When biomass species are reproduced again by photosynthesis, the cycle repeats (Figure 12.1). For a steady operation, the concentration of CO_2 in the atmosphere remains constant. However, the heavy use of fossil fuels, which is not part of the regular carbon cycle, has caused increase in the atmospheric CO_2 concentration.

The conversion of biomass into energy is receiving increasing interest due to the need for maintaining the carbon cycle, and the depletion of petroleum fuels. Biomass is likely to play an important role in limiting CO_2 emissions while supplying energy needs related to electric power, transportation fuels, and heating. Compared with the redesign of fossil fuel processes for lower CO_2 emissions, biomass has a higher effectiveness in reducing CO_2 emissions due to potential net zero carbon emission considering the production/use cycle (Hall, Mynick, and Williams, 1991; Overend, 1996; Williams and Larson, 1996). Given the growing shortage of petroleum fuels and environmental pollution (especially the greenhouse gas effect), the need for developing a carbon-neutral renewable source of energy is greater now than ever before (Lal, 2005). In this respect, biomass is likely to become a key source of renewable energy for the sustainable society of the future.

12.2 Environmental Impacts of Biomass Production

Biomass for energy crops include fast-growing trees (e.g., hybrid poplar, black locust, willow, silver maple), annual crops (e.g., corn, sweet sorghum), and perennial grasses (e.g., switchgrass, miscanthus). Carbon and energy from biomass can be converted into modern energy carriers, such as gasoline, diesel, ethanol, hydrogen,

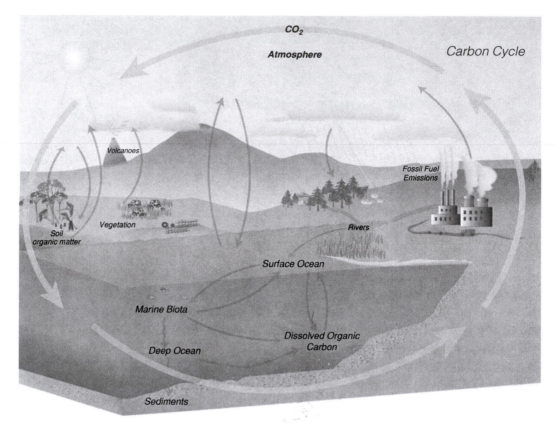

Figure 12.1. The carbon cycle (ESRL, 2009).

and electricity. If a large amount of land is removed from food production and put into biomass production, the environmental impacts can be substantial if not properly managed. Hence, any negative effects of biomass production on erosion, water quality, chemical and fertilizer fates, wildlife habitat, and biodiversity should be eliminated or minimized.

12.2.1 Land Use

The first-generation biofuels, such as ethanol from corn, wheat, or sugar beets, and biodiesel from oil seeds, are produced from classic agricultural food crops that require high-quality agricultural land for growth. On the contrary, the second-generation biofuels (i.e., lignocellulosics nonfood biomass) can use marginal-quality land. It is perceived that some of the nature-land that has been kept out of human activity may be used. A basic question arises about whether there is enough land to produce biofuel at a large scale without disturbing its traditional use for food, feed, and fiber production.

A 2005 study attempted to determine whether there is enough land to produce 1.3 billion tons/year dry biomass needed to displace about 30% of U.S. liquid fuel used in transportation (ORNL, 2005). The report of this study projects that

1.3 billion dry tons per year of biomass could be available for large-scale biofuel industry by the mid-twenty-first century, while still meeting demands for food, fiber, and forestry products. This supply of biomass would be a sevenfold increase over the 0.19 billion dry tons of biomass per year currently used for bioenergy and bio-products. Land area in the United States is about 2 billion acres, with 33% forest-lands and 46% agricultural lands (26% grasslands or pasture and 20% croplands). The analysis excluded environmentally sensitive areas, lands without road access, regions reserved for nontimber uses (e.g., parks and wilderness), and lands used permanently for pasture. A total of 448 million acres of agricultural lands (both active and idle) were included in this study. This study assumed that 55 million acres of idle cropland would be dedicated to the production of perennial bioenergy crops, and all other cropland would continue to be used for traditional food, feed, and fiber crops. Harvesting agricultural residues (e.g., stems, stalks, and leaves that are not currently used) from these traditional crops would be a significant source of biomass.

The amount of land used for global agriculture and goods production is about 3.7 billion acres, which represents about 11% of Earth's surface. Additionally, there is the potential for 6.9 billion acres for use of biomass production. However, part of the potential land is devoted to other uses: 45% for wood and forest, 12% for protected area, and 3% for human settlement. Current farmland use for biofuels is only 34.6 million acres, which is approximately 1% of the entire cultivated land in the world (Escobar et al., 2009)

12.2.2 Irrigation Water Consumption

Environmental impacts of irrigation are the changes in quantity and quality of water and soil as a result of the irrigation and the ensuing effects on downstream natural and social conditions. Irrigation will be an essential component of any strategy to increase the global food supply and biofuel crop production. Currently, the irrigated agriculture depends on supplies from surface or ground water. With an increase in biomass production, the consumption of irrigation water will sharply increase. Humans are now faced with the question of how to increase the agricultural irrigation given current global shortages of water and the conflict between food supply and energy demand. Improving the environmental performance of current and future irrigation is important for long-term sustainability. The environmental impact depends on the nature of the water source, the quality of the water, and how water is delivered to the irrigated land. In addition, the management of water application systems as well as the suitability of related agronomic practices will have a dramatic influence on the environmental impact.

12.2.3 Fertilizer and Pesticide Use

Some of the biomass crops have been claimed to grow without much fertilizer or pesticide use. But, in order to obtain high yields for a commercially viable operation, it is likely that fertilizer and pesticide use will increase dramatically. However, some of the biofuel process may be able to return some fertilizer back to the soil (e.g., biochar, ash-containing nutrients). Surface runoff carries manure, fertilizers, and pesticides into streams, lakes, and reservoirs, in some cases causing unacceptable levels of bacteria, nutrients, or synthetic organic compounds. Hence, proper assessment needs to be made. The soil erosion resulting from intensive farmland can cause a number of environment problems. For example, erosion in U.S. agricultural fields is about tenfold more than the natural erosion. Erosion can affect the biomass productivity as it removes the most productive surface soil. The remaining subsoil is usually less fertile, less water-absorbent, and less able to retain pesticides and fertilizers. With intensive biomass production, erosion may increase. On the other hand, some perennial biomass crops (e.g., switchgrass) can provide an extensive root system to keep soil from eroding.

12.2.4 Ecosystem Diversity

Biomass production from a few crops that have been shown to give the highest yield can impact the biodiversity in nature. For economic reasons, farmers are likely to plant uniform energy crops consistent with conventional agricultural practice. This will reduce the number of diverse plant species. Once extinct, the species will never return and will not be available for maintenance of a complex ecological network. Also, the use of a few monoculture crops increases the global risk on food and fuel supply in case of natural catastrophes. However, with judicious isolation of land areas, it may be possible to maintain some diversity in each of the sub areas.

12.3 Environmental Impacts of Biomass-to-Biofuel Conversion

Through the biological and thermochemical conversion, biomass can be converted into liquid biofuels. At present, due to lack of commercial-level plants, environmental impacts of many of the processes involved have not been fully established. In the case of first-generation biofuels (ethanol and biodiesel), concerns have been due to the discharge of liquid effluents. Some of the very small-level plant operators have not complied with the discharge of methanol- and glycerin-containing effluents. In the case of second-generation biofuel, biomass gasification and pyrolysis processes have met with some resistance from the environmental community and the public. In these processes, proper design of exhaust emission control systems is necessary to ensure that health and safety requirements are met. The gaseous effluent from pyrolysis and gasification reactors can contain a variety of air pollutants that must be controlled prior to discharge into the ambient air. These include particulate matter, aerosols or tars, nitrogen oxides, sulfur oxides, dioxins and furans, hydrocarbon gases, multiple metals, and carbon monoxide. From experience with coal-based power plants, there exist a number of strategies for controlling emissions that can be used in the biomass gasification and pyrolysis operations. However, these would be highly dependent on the process requirements of each individual facility.

Table 12.1. *Emission impacts of adding 20 vol% soybean-based biodiesel to petroleum diesel (U.S. EPA, 2002)*

	Percent change in emissions
Nitrogen oxides (NO_x)	+2.0
Particulate matter	−10.1
Hydrocarbons	−21.1
Carbon monoxide (CO)	−11.0

Use of process water during biomass-to-biofuel conversion is another concern. On average, ethanol corn-to-ethanol plant consumes about 4 gallons of water per gallon of ethanol produced, which is much higher than water use of 1.5 gallons/gallon fuel in the petroleum industry. In the case of cellulosic ethanol, due to additional processing steps needed, this industry may require as high as 9 gallons of water for each gallon of ethanol produced.

12.4 Environmental Impacts of Biofuel Use

The biggest difference between biofuels and petroleum fuels is the oxygen content. Non-upgraded biofuels have 10–45 wt% oxygen, whereas petroleum fuels have less than 1%, making the combustion properties of biofuels very different from petroleum. There are additional differences; for example, biodiesel is different than conventional diesel in terms of its sulfur content, aromatic content, and flash point. It is essentially a sulfur-free and nonaromatic compound, whereas petroleum diesel contains up to 500 ppm sulfur and 20–40 wt% aromatic compounds. Although aromatic content can swell the elastomers and provide a good seal in the fuel injection system, only a few percent aromatic is enough to provide this benefit. Biodiesel advantages could be a key solution to reducing the problem of urban pollution since gas emissions from the transportation sector is significant. Diesel, in particular, is dominant for black smoke particulate together with sulfur dioxide (SO_2) emissions and contributes to a one-third of the total transport-generated greenhouse gas emissions (Nas and Berktay, 2007). By using biodiesel (B100), one can decrease emissions on average of 14% for CO_2, 17.1% for CO, and 22.5% for smoke density (Utlu, 2007).

Blends of up to 20 wt% biodiesel mixed with petroleum diesel can be used in nearly all diesel equipment and are compatible with most storage and distribution equipment. Using biodiesel substantially reduces emissions of unburned hydrocarbons, carbon monoxide, sulfates, polycyclic aromatic hydrocarbons, nitrated polycyclic aromatic hydrocarbons, and particulate matter. The emission performance improves as the amount of biodiesel blended into diesel fuel increases. An exception is NO_x emission, which increases with biodiesel use (Demirbas, 2009). Some of the emission impacts of adding biodiesel to petroleum diesel are shown in Table 12.1.

Blending ethanol with gasoline can significantly change the emissions (Graham, Belisle, and Baas, 2008). For example, 10 vol% ethanol blending causes a decrease in CO emission (-16%); increases in emissions of acetaldehyde (108%), 1,3-butadiene (16%), and benzene (15%); and no significant changes in NO_x, CO_2, CH_4, N_2O, or formaldehyde emissions. On the other hand, blending 85% ethanol results in decreases in emissions of NO_x (-45%), 1,3-butadiene (-77%), and benzene (-76%); increases in emissions of formaldehyde (73%) and acetaldehyde (25–40%), and no change in CO and CO_2 emissions.

Blending of ethanol with diesel can potentially reduce the emission of particulate matter. For example, 10% ethanol addition can decrease the total number and total mass of the particulate matters by about 12–27% (Kim and Choi, 2008). However, the disadvantages of ethanol include its lower energy density, corrosiveness, low flame luminosity, lower vapor pressure, miscibility with water, and toxicity to ecosystems.

SO_2 emission from the combustion sulfur in fossil fuels causes acid rain, which has been shown to have adverse effects on forests, freshwaters, and soils, killing insect and aquatic lifeforms as well as causing damage to buildings. In recent years, many governments have introduced laws to reduce SO_2 emissions, for example, by capturing SO_2 from power plant exhaust and mandating ultra-low-sulfur diesel. Biofuel contains virtually trace amounts of sulfur, so its blending with petroleum fuels can reduce SO_2 emissions. In addition, solid biomass can be directly used with coal in the power plants.

12.5 Life-Cycle Impacts

Although the environmental benefits of ethanol use have been debated, the benefits of second-generation biofuels have a clear positive environmental impact, mainly due to the CO_2-neutral nature of the fuel. This is because growing biomass absorbs CO_2 similar to the amount emitted when biofuel is burned. The benefit may be slightly diminished if one considers the use of fossil fuel in collection and transportation of the biomass (Petrou and Pappis, 2009). However, if the whole biomass seed-to-liquid-fuel chain operation is performed using biofuel itself, then some of the benefits are recovered, except those lost to the use of fossil fuel to produce machinery and fertilizers. A rigorous analysis of all energy inputs to biofuel production should be performed.

A useful tool to determine the overall environmental impact is the life cycle analysis (LCA) of biofuel, in which all inputs and outputs starting from the biomass growth to the final biofuel use are considered (Escobar et al., 2009). Unfortunately, the literature is full of many conflicting biofuel LCAs, mostly due to assigning varying values to crop cultivation, crop transport, and biomass-to-biofuel conversion. Additional variations arise due to the varying choices of system boundaries. Several indicators to evaluate the overall environmental and energy performance of biofuel have been used as described below (Davis, Anderson-Teixeira, and DeLucia, 2009). Net energy value (NEV) is a commonly reported energy efficiency term defined as

$$NEV = \text{(usable energy produced from a biofuel)}$$
$$- \text{(amount of energy required in the production of the biofuel)}$$

Table 12.2. *Reported NEV of biofuel crops from LCA (Davis et al., 2009)*

Biofuel crop	NEVs (MJ/m^2 of land used)		
	Maximum reported	Minimum reported	Most often reported
Corn	2.3	−2.52	1.6
Switchgrass	7.0	−2.6	6.0
General lignocellulosic crop			4.5
Reed canary grass			4.9

A negative NEV represents a net energy loss, where more energy is required to produce the biofuel than the fuel can provide. A positive NEV represents a net energy gain, where the biofuel can provide more energy than it consumes during its production. Reported range of NEV values for various biofuel crops are shown in Table 12.2.

The second-generation biofuel crops (switchgrass and Reed canary grass) have a much better NEV then the first-generation biofuel crop (corn). In the case of switchgrass, each m^2 land can produce net energy of 6.0 MJ every year, or 100 m × 100 m (or 1 hectare, or 2.47 acres) land can produce 60 GJ of net energy.

Fuel energy ratio (FER) is another way of reporting fuel efficiency, defined as

$$FER = \text{(amount of biofuel energy produced)}/\text{(amount of fossil energy required to manufacture the biofuel)}$$

A FER value of greater than unity represents a net energy gain, whereas a FER value less than unity represents net energy loss. Published FER (Table 12.3) values for a given biofuel crop show a large variation.

Greenhouse gas (GHG) displacement is another measure that indicates the net change in the GHG emission by substituting fossil fuel with biofuel. Hence, a negative value indicates a reduction in the GHG emission and a positive value indicates an increase in the GHG emission, as compared with the base case of continuing to use fossil fuel. Various estimated GHGs from LCAs are shown in Table 12.4.

Clearly, the displacement of fossil fuel by second-generation biofuel causes a significant decrease in GHG emissions, except for a few studies that show a net increase in the GHG emission.

Table 12.3. *Reported FER for biofuel crops from LCA (Davis et al., 2009)*

Biofuel crop	FER (MJ/MJ)		
	Maximum reported	Minimum reported	Most often reported
Corn	2.0	0.7	1.5
Switchgrass	4.4	0.44	4.4
General lignocellulosic crop	5.6	1.8	4.3
Miscanthus	1.2	1.0	1.2

Table 12.4. *Reported GHG displacements by biofuel crops from LCA (Davis et al., 2009)*

Biofuel crop	GHG displacement (%, base case is fossil fuel use)		
	Maximum reported	Minimum reported	Most often reported
Corn	−86	93	−25
Switchgrass	−114	43	−73
General lignocellulosic crop			−80
Miscanthus			−98
Reed canary grass			−84

12.6 Summary

Urban air quality is likely to improve with the substitution of petroleum fuels with biofuels. The global environmental impact of second-generation biofuels is much more favorable than first-generation biofuels. Various LCAs of biofuel crops have been reported, unfortunately with widely varying results. Nonetheless, a majority of studies indicate that the use of biofuels can reduce both GHG emission and dependence on petroleum fuels. However, there are some concerns due to impact on land use, irrigation, and biodiversity. Careful planning can alleviate some of the concerns. For example, the planners should carefully treat the relationship between the land, irrigation, fertilizer and pesticide use, and ecosystem diversity when planning for large-scale biomass production to substantially reduce the use of fossil fuels.

13 Economic Impact of Biofuels

13.1 Biofuel Economy

High petroleum prices, increasing trade deficits due to fuel import, depleting petroleum supplies, and increased concern over the greenhouse gas emissions from fossil fuels have driven interest in transportation biofuels (Hill et al., 2006). For many developing countries, petroleum import accounts for the major share of their outflow of foreign exchange, putting a high strain on the growth. For such countries, a locally produced fuel can help with the economy. The first-generation biofuels (ethanol and biodiesel) are already in the market, and the second-generation biofuels from biomass are emerging. The U.S. Energy Independence and Security Act of 2007 provides major support for the biofuel industry by setting a renewable fuel requirement for use of 36 billion gallons per year of biofuel by 2022, with corn ethanol limited to 15 billion gallons. Any other ethanol or biodiesel may be used to fulfill the balance of the mandate, but the balance must include 16 billion gallons per year of cellulosic ethanol by 2022 and 5 billion gallons per year of biodiesel by 2012.

With increasing global interest in biofuels, there is a considerable discussion on the biomass supply and conversion costs. Production and use of life cycle analysis are being carried out to determine whether biofuels provide any benefit over fossil fuels, accounting for various aspects such as farm yields, commodity and fuel prices, farm energy and agrichemical inputs, production plant efficiencies, coproducts, greenhouse gas emissions, and so on. The biofuel industry does present an excellent opportunity for economic development of rural areas (Leistritz and Hodur, 2008). In general, first-generation biofuels are more expensive then petroleum fuels when wholesale prices are compared at the same energy content basis. For illustration, the wholesale U.S. prices for fuels in 2008 and 2009 are shown in Table 13.1. Typically, 1 GJ of energy is needed to drive 250 km. Even when crude oil prices peaked in 2008, per GJ price of ethanol or biodiesel was higher than gasoline or diesel by about 25%. When the crude oil prices dropped significantly during the economic recession of 2009, ethanol and biodiesel became almost twice as expensive as gasoline or diesel. But the economic impact is not fully described by price alone. As

Table 13.1. *Wholesale prices for fuels in 2008 and 2009 in the United States*

Fuel	Heating value (lower)	July 2008 prices	$/GJ	March 2009 prices	$/GJ
Gasoline	121.7 MJ/gallon	$3.1/gallon	25.2	$1.3/gallon	10.7
Jet fuel	116.8 MJ/gallon	$2.9/gallon	24.6	$1.2/gallon	10.3
Diesel	126.0 MJ/gallon	$4.2/gallon	33.3	$1.2/gallon	9.5
Natural gas	38.1 MJ/kg	$9.2/Million BTU	8.7	$4.3/million BTU	4.1
Heating oil	133.6 MJ/gallon	$3.6/gallon	26.6	$1.2/gallon	9.0
Coal	33.3 MJ/kg	$100/ton	3.0	$66/ton	2.0
Soybean oil	32.5 MJ/kg	$0.6/lb	40.8	$0.3/lb	20.4
Corn	15 MJ/kg	$5.8/bushel; $0.23/kg	15.3	$3.4/bushel; $0.13/kg	8.9
Ethanol	79.8 MJ/gallon	$2.5/gallon	31.1	$1.6/gallon	20.1
Methanol	18.0 MJ/kg	$1.4/gallon or $526/ton	29.2	$0.65/gallon or $216/ton	13.8
Biodiesel	123.5 GJ/gallon	$5.5/gallon	44.5	$2.6/gallon	21.0
Wood (biomass)	15 MJ/kg	$100/dry-ton ($50/green ton)	6.6	$50/ton ($25/green ton)	3.3

most of the biofuels are locally grown, the impact on local economies has been significant. One of the key reasons for the cost of ethanol and biodiesel is the feedstock cost (grains, vegetable oil). In the case of second-generation biofuels from biomass, the feedstock is much cheaper.

13.2 Economic Impact of Corn Ethanol

Recently, the ethanol industry in the United States has undergone a rapid growth. For example, the number of plants in the United States has increased from 54 in 2000 to about 200 in 2008, with a combined installed capacity of 13 billion gallons per year (RFA, 2009). The growth is mostly concentrated in Iowa, Nebraska, Illinois, Minnesota, South Dakota, Indiana, Michigan, and Ohio, in decreasing order. Estimates of the impact on local economies vary wildly, mainly because of assumptions regarding the corn industry (Low and Isserman, 2009).

Economic impacts are realized through all inputs and outputs related to the ethanol plant. The main inputs are corn, natural gas, chemicals, yeast, enzymes, water, and labor (Low and Isserman, 2009). Corn is generally purchased directly from farmers, and then transported by trucks within a 50-mile radius of the plant. About half of the input price is from corn, natural gas, and electricity; hence, ethanol plant operation is very sensitive to these inputs. Tyner and Taheripour (2007) estimated a break-even per-bushel corn price for a new plant of roughly $3.70/bushel when crude oil is $40/barrel and $4.72/bushel when crude oil is $60/barrel. But without the $0.51/gallon U.S. subsidy for ethanol, break-even corn price is only $3.12/bushel when crude oil is $60/barrel (i.e., corn price has to be less than $3.12/bushel for ethanol plant to be profitable). The produced ethanol is transported (usually by trucks) to gasoline suppliers. Ninety percent of plant revenue comes from the sale of ethanol and the remaining 10% from byproduct cattle feed

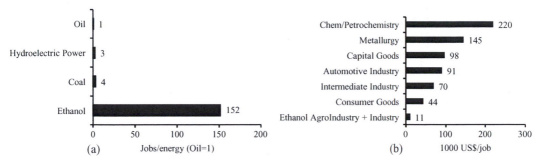

Figure 13.1. Employment comparison (a) and investment for job creation (b) in various industries in Brazil (adapted from Goldemberg, 2002).

(Low and Isserman, 2009). Typically a new ethanol plant (e.g, 50 million gallons ethanol/year capacity) directly employs 20–50 employees. But employment count is as high as 133 jobs when indirect employment is included. In addition, the total economic activity is of the order of $140 million/year (Swenson 2008). Hence, for a 13-billion-gallon/year operation, a rough estimate on economic impact is $36 billion with employment for about 35,000 people.

13.3 Economic Impact of Sugarcane Ethanol

Ethanol from sugarcane, produced mainly in developing countries with warm climates, is generally much cheaper to produce than ethanol from grain or sugar beet in developed countries. The ethanol from sugarcane has a much more favorable energy output/input ratio as compared with corn (about 9 for sugarcane versus 1.5 for corn). For this reason, in countries like Brazil and India where sugarcane is produced in substantial volumes, sugarcane-based ethanol is becoming an increasingly cost-effective alternative to petroleum fuels. Hence, ethanol from the sugarcane industry has had a significant impact on the local economies (Goldemberg, Coelho, and Guardabassi, 2008). Currently, Brazil produces about 6.5 billion gallons of ethanol per year. According to the Brazilian Ethanol Program, Proalcool, the number of jobs created by given investment is highest in the ethanol agricultural industry (Figure 13.1b). In addition, out of various energy options, per unit energy basis, the ethanol industry employs 152 times more people than the petroleum industry, and 38 times more people than the coal industry (Figure 13.1a).

Only sugarcane ethanol comes close to competing with gasoline. Comparatively, corn ethanol is considerably more expensive, and ethanol from other grains and sugar beet in Europe is even more expensive. These differences arise due to many factors, including feedstock cost, scale of operation, process efficiency, capital and labor costs, coproduct revenues, and government subsidies. Estimates show that ethanol from sugarcane, corn, or sugar beet becomes cost-competitive when crude oil prices are $20–30, $50–60, or $70 per barrel, respectively (Dufey, 2006; Balat, 2007).

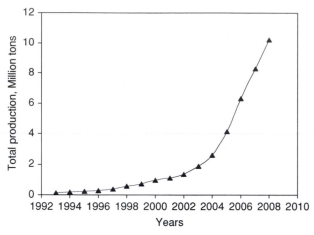

Figure 13.2. World production of biodiesel, 1980–2008 (adapted from Demirbas, 2008b).

13.4 Economic Impact of Biodiesel

For the production of biodiesel from vegetable oils, technology and equipment requirements are simple. That is why biodiesel production has developed almost like a cottage industry in many countries, with small-scale production plants in rural areas. In countries where diesel is heavily taxed, some farmers have started producing biodiesel for in-house use for fueling the farm machinery. This way, farm-produced vegetable oil is directly used and external diesel purchase is avoided. Over the last five years, worldwide biodiesel production has grown rapidly (Figure 13.2).

The cost of feedstock (i.e., vegetable oil, methanol) is a major factor in the viability of commercial biodiesel production. For example, according to the estimate of Hass et al. (2006), the overall cost of soy-based biodiesel is $0.70/liter (or $2.64/gallon), which is composed of $0.54/liter (or $2.04/gallon) feedstock cost and $0.16/liter (or $0.60/gallon) processing cost. Hence, as the vegetable oil and methanol prices fluctuate due to the market conditions, the cost of biodiesel manufacturing follows. In recent years, the price of edible oil has gone up sharply due to increased demand of edible oil for biodiesel manufacturing (see Figure 7.2 in Chapter 7). The competition of edible oil sources as food versus fuel makes edible oil not an ideal feedstock for biodiesel production (Gui, Lee, and Bhatia, 2008). The focus should be on using the waste oil or nonedible oil for biodiesel manufacturing. Farming of nonedible oil (e.g., jatropha) in marginal lands can give rise to rural employment, because such operations are highly labor-intensive. Biodiesel from waste oil and grease from the food industry is an excellent option as this feedstock can be purchased at much lower prices then the original vegetable oil, and it also solves the environmental disposal problem of waste oil.

13.5 Future Economic Impact of Biomass-Based Biofuels

Unfortunately, corn ethanol or biodiesel cannot replace petroleum without impacting food supplies. However, a substantial impact can be realized from the use of

Table 13.2. *Projected ethanol production in the United States (De La Torre Ugarte et al., 2007)*

Feedstock	Billion gallons of ethanol		
	Year 2010	Year 2020	Year 2030
Corn grain	10.00	14.09	14.09
Wood residue	0.00	2.33	4.54
Wheat straw	0.00	0.01	1.14
Corn stover	0.00	0.00	8.88
Dedicated energy crops	0.00	13.69	31.71
Total ethanol production	10.00	30.11	60.35

biofuels derived from biomass. The low-input biomass grown on agriculturally marginal land or from waste biomass could provide much greater economic and environmental benefits than food-based biofuels (Hill et al., 2006).

Recently De La Torre Ugarte, English, and Jensen (2007) carried out an economic impact analysis for producing 60 billion gallons/year ethanol and 1.6 billion gallons of biodiesel by 2030 in the United States. About 23% ethanol is derived from corn and the remaining 77% from biomass (Table 13.2), assuming that cellulose-to-ethanol technology will be commercial by 2012. The estimates show that, by 2030, dedicated energy crop areas will increase from near zero to 34.4 million acres, pasture land will decrease from 56.7 to 24.3 million acres, and corn crop areas will increase from 81 to 83 million acres. The biofuel crops are expected to have some effect on the major crop prices. Changes in these prices in 2006 are shown in Table 13.3.

During the time period 2007–2030, the projected cumulative increase in net farm income is estimated to be $210 billion. The direct economic impact is expected to be $110 billion/year (using year 2006 dollar value) due to the purchase of inputs, value addition, and sale of biofuel. The economic impact will be shared by biofuel (23%) and agriculture (77%) industries, resulting in the creation of 58,000 jobs in the biofuel industry and 236,000 jobs in the agriculture industry. But, when all of the indirect impacts are accounted for, estimated total economic impact is expected to be $368 billion with the creation of 2.4 million jobs.

The economic change will be felt more in the areas that produce biomass. A scenario of biomass production for geographic areas in 2030 is shown in Figure 13.3; this

Table 13.3. *Impact of biofuel production on the major crop prices in the United States (De La Torre Ugarte et al., 2007)*

Crop	Change in crop prices from year 2006		
	Year 2010	Year 2020	Year 2030
Corn (Δ$/bushel)	0.89	0.36	0.62
Wheat (Δ$/bushel)	0.11	0.00	0.36
Soybeans (Δ$/bushel)	0.82	0.97	1.23
Cotton (Δ$/pound)	0.00	0.03	0.02

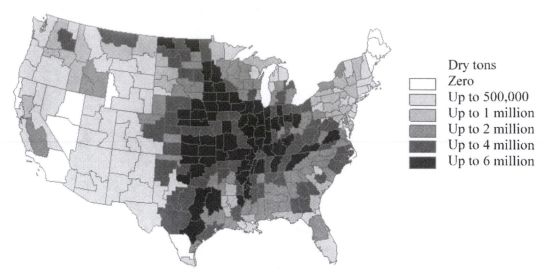

Figure 13.3. Projected U.S. biomass production in 2030 (De La Torre Ugarte et al., 2007; reproduced with permission of Wiley).

shows that a wide area of the United States will contribute to the biofuel industry. The production of dedicated energy crops is concentrated in the Southeast, Southern Plains, and Northern Plains, whereas corn stover is concentrated in the Midwest (De La Torre Ugarte et al., 2007).

The location of biofuel facilities will have proximity to the biomass production as biomass cost increases with transportation distance as shown in Figure 13.4 for the case of Mississippi. In addition, biomass cost increases with its demand. Grebner et al. (2008) note that 4 million dry tons of woody biomass can be produced in Mississippi alone every year, consisting of logging residue, small-diameter

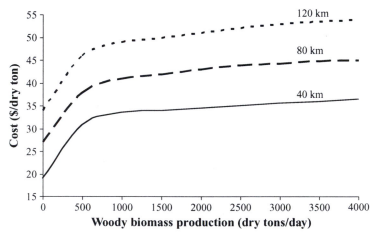

Figure 13.4. Cost of woody biomass feedstock as a function of distance and demand volume in Mississippi (adapted from Grebner et al., 2008; x-axis was modified from dry tons to dry tons/day).

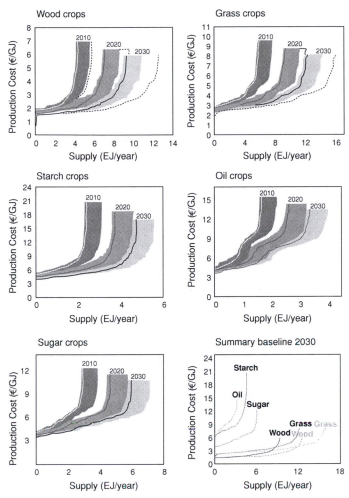

Figure 13.5. Cost–supply curves for five assessed crop groups. The summary figure depicts cost–supply for all crop groups for the 2030 curves for the baseline scenario (De Wit and Faaij, 2009; reproduced with permission from Elsevier).

trees, mill residue, and urban waste. Small-diameter tree biomass can be produced at $49/dry ton, whereas logging residue (being a byproduct) can be obtained at $40/dry ton from a 50-mile procurement radius (Perez-Verdin et al., 2008).

De Wit and Faaij (2009) assessed the cost and supply of potential biomass resources in Europe. It is suggested that, by 2030, total land of 900,000 km^2 (about 222 million acres) could be available for biomass production. Supply and production cost curves for five types of crops (wood, grass, starch, oil, and sugar) are presented in Figure 13.5. Oil, starch, and sugar crops will have limited supplies as the cost increases significantly even with a lower production volume. On the other hand, wood and grass crops can provide significantly more biomass. In combination, 27.7 EJ/year supplies are expected by 2030, which is equivalent to 1.5 billion tons/year of dry lignocellulosic biomass.

13.6 Economic Impact on Developing and Rural Economies

In developed countries, the cost of large-scale biofuel production is currently high. For example, the production cost of biofuels may be three times higher than that of petroleum fuels, without, however, considering the nonmarket benefits. Conversely, in developing countries, the costs of producing biofuels are much lower than in the OECD countries and very near to the pretax price of petroleum fuel in those countries (United Nations, 2006). Hence, more sustainable impact of biofuels can be realized in developing countries.

The economy in developing countries is based on agricultural production, and most people live in rural areas to support the agriculture. The rural population can be further employed to produce biomass. For such countries, the economic advantages of a biofuels industry would include value added to the feedstock, an increased number of rural manufacturing jobs, increased income taxes, investments in plant and equipment, reduced greenhouse gas emissions, reducing a country's reliance on crude oil imports, and supporting agriculture by providing new labor and market opportunities for domestic crops. In fact, green energy production may become a principal contributor in economic development of a developing country. However, its implementation should be integrated with whole community development programs.

The biofuel industry may provide a means to alleviate rural poverty by providing more economic activity in those areas. In developing countries, poverty in rural areas and the lack of programs and funding for agricultural development are the most important causes of nourishment insecurity, conflicts, terrorism, corruption, and environmental degradation. Food production in the world has increased substantially, but there is a significant population that suffers from malnourishment. In fact, the majority of the suffering population depends on local agriculture, which can be helped with rural development. Thus, countries with suitable climatic and land potential for the production of biofuels have a significant opportunity for developing their agricultural regions, which can improve the population's standard of living and income. On the other hand, considering that the amount of land in the world available for agriculture is limited, it is necessary to define the fraction of farmland that could be used for the production of biofuels.

With the production of local fuel, countries will be less affected by the global fluctuations in the petroleum supply and prices. In addition, local biofuels can provide a security against possible disruptions in the petroleum imports due to natural disasters or political reasons. Hence, biofuels come with the security of supply benefits that should be considered in the overall evaluation.

13.7 Summary

First-generation biofuels (ethanol and biodiesel) in developed countries are not cost-competitive when compared with petroleum. However, in some of the developing countries, the input costs are lower and these fuels provide a significant positive economic impact. Nonetheless, local economies do benefit from biofuel industries in all countries due to economic activity and job creation. Forecasts on second-generation biofuels from biomass show a very positive overall economic impact with a significant employment creation opportunity. In addition, the biofuel industry offers an opportunity to address rural poverty by providing economic activity in rural areas.

Biofuel Policy

14.1 Introduction

Energy policy deals with issues related to the production, distribution, and consumption of energy. It is the manner by which a government/entity has decided to address these issues, including international treaties, legislation on commercial energy activities (trading, transport, storage, etc.), incentives to investment, guidelines for energy production, conversion and use (efficiency and emission standards), taxation and other public policy techniques, energy-related research and development, energy diversity, and avoidance of possible energy crises. Current energy policies also address environmental issues, including cleaner technologies, more efficient energy use, air pollution, greenhouse effect (mainly reducing carbon dioxide emissions), global warming, and climate change (Demirbas, 2007a, 2008a, 2008c).

For commercial biofuel operations, there is considerable uncertainty in the economic analysis, mostly associated with the costs of feedstock, rapid technological changes, and petroleum prices against which the biofuels have to compete. Because the biofuel industry is in its infancy, supportive policies are needed that can foster the initial growth for both first-generation (ethanol and diesel) and second-generation biofuels. Policy options include incentive payments or tax breaks. Due to rising prices of petroleum fuels, the competitiveness of biomass use has improved considerably over time. Biomass and bioenergy are now a key option in energy policies for several countries, mainly driven by the security of supply, reducing both petroleum import and carbon dioxide (CO_2) emissions. Biofuel can provide new economic opportunities for people in rural areas of countries with biofuel potential. Targets and expectations for bioenergy in many national policies are ambitious, reaching 20–30% of total energy demand in various countries (Table 14.1)

Many governments are actively supporting production of liquid biofuels in their countries by using a number of policy mechanisms, including minimum price guarantee, tax incentives, guaranteed markets, and other direct support for bioenergy production, such as grants, loans, and subsidies (GBEP, 2007). Government policies are therefore playing a key role in influencing investment in bioenergy area. Following GBEP (2007), there are four main factors driving the current interest in

Table 14.1. *Biofuel targets for various countries*

Country	Biofuel target
Europe	5.75% by 2010
France	7% by 2010 and 10% by 2015
United Kingdom	5% of transportation fuel by 2010
Australia	350 million liters
Canada	5% of gasoline by 2010; 2% of diesel and heating oil by 2012
New Zealand	3.4% of transportation fuel by 2012
United States	36 billion gallons (or 136 billion liters) by 2022
Brazil	5% of diesel by 2013; 20–25% of gasoline presently
China	2.2 million tons by 2010 and 12 million tons (or 15 billion liters) by 2020
India	20% by 2017
Turkey	5% by 2020

liquid biofuel: (1) high petroleum prices, (2) energy security, (3) climate change, and (4) rural development. Policies of some of the specific countries are discussed below.

14.2 Brazilian Biofuel Policy

Brazil, a vast country with abundant natural resources and agricultural land, has emerged as a global leader in the production of ethanol from sugarcane. Brazil is the largest single producer of sugarcane with about 27% of global production and a high yield of 7.3 dry tons/acre (Kim and Dale, 2004). J. W. Bautista Vidal (the former Secretary of Science and Technology accredited for the creation of the Brazilian National Fuel Alcohol Program, PROALCOOL, in 1975) believed that Brazil is in a very special and historic position to transform itself into an important force in the production of renewable liquid energy, as a leader in a new "civilization of photosynthesis," contributing to the preservation of a peaceful energy supply (Vidal, 2006). In addition, the Brazilian Agro-Energy Plan 2006–2011 (prepared by the Ministry of Agriculture, Livestock, and Food Supply) considered that Brazil possesses a series of advantages that will propel the country into a leadership position in the global biofuel market. The advantages include the availability of agricultural land, a tropical climate, abundant biodiversity and water resources, an established agricultural industry, and a large domestic biofuel market to support the international expansion (Oliveira and Ramalho, 2006).

Brazil initiated large-scale ethanol production in the late 1970s after a 1975 law guaranteed price parity for ethanol and gasoline along with a system of tax rebates and subsidies for ethanol plants (Walter and Cortez, 1999; Colares, 2008). In addition, since 1979, the Brazilian sugar and ethanol industry has invested approximately $40 million/year in research and development. The amount of alcohol blended into gasoline is dictated to the market by law or decree, which directly affects the Brazilian producer prices of sugarcane, consumer prices for sugar and ethanol, and sugar quantities both produced and consumed in Brazil, as well as world prices for sugar

(Schmitz, Seale, and Buzzanell, 2002). About one-half of the sugarcane produced in Brazil is used for ethanol production, which has no government limits on production. Typically, ethanol is produced both at sugarcane mills with adjoining distillery plants, producing both sugar and ethanol, and at independent distilleries, producing only ethanol.

Brazil's ethanol industry has shown that it is possible to be efficient, both environmentally and economically, in the implementation of a biofuel infrastructure. On the other hand, the North American expansion of corn ethanol has required subsidies and trade barriers, as ethanol is more expansive to produce from corn than from sugarcane. However, it should be noted that sugarcane-to-ethanol processes have been used on a large scale in Brazil for more than 25 years and are much more mature than the processes using corn. The optimization during this long duration has resulted in major reductions in the production costs (Goldemberg et al., 2004).

For biodiesel, in 2005, Brazil added a 2% blending mandate by 2008, which was increased to 5% by 2013 (Colares, 2008; OECD, 2008). The Brazilian Biodiesel Policy does possess the laudable objectives of regional development and promoting social inclusion. However, the majority of authorized biodiesel production capacity being installed in the Central-West of the country is mainly attributed to the lack of strong policy mechanisms to spread the production throughout the country. The role of family farmer is being limited only to producers of soybeans. The biodiesel production permits are being auctioned. For example, 1,920,000 m^3 of biodiesel were auctioned with an average price of R$2137.04/m^3 during 2005–2008.

14.3 European Biofuel Policy

In recent years, convergence of different economic and environmental forces has generated interest in renewable energy in the European Union (EU). During the 1990s, the production and use of ethanol and biodiesel started in several European countries and expended significantly. The European Commission (EC) is using both legislation and formal directives to promote biofuel production and use within the EU. The motivation has been from various policy goals, such as reducing greenhouse gas emissions, boosting the decarbonization of transport fuels, diversifying fuel supply sources, and developing long-term replacements for petroleum while diversifying income and employment in rural areas. The general EU policy objectives most relevant to the design of energy policy are (1) competitiveness of the EU economy, (2) security of energy supply, and (3) environmental protection (Jansen, 2003). The key elements of the European biofuel economic policy (EC, 2003; Jansen, 2003; Hansen, Zhang, and Lyne, 2005) are as follows:

- A Communication presenting the action plan for the promotion of biofuel and other alternative fuels in road transport.
- The Directive on the promotion of biofuel for transport, which requires an increasing proportion of biofuel in all diesel and gasoline sold in the Member States.

Table 14.2. *Percent shares of alternative fuels in transportation in the EU under the optimistic development scenario of the EC*

Year	Biofuel	Natural gas	Hydrogen	Total
2010	6	2	–	8
2015	7	5	2	14
2020	8	10	5	23

- The biofuel taxation, which is part of the large draft Directive on the taxation of energy products and electricity, proposing to allow Member States to apply differentiated tax rates in favor of biofuel.

The EU have also adopted a directive for the use of biofuel blending in gasoline and diesel, with minimum content of at least 2% by 2005, and then increasing in stages to a minimum of 5.75% by 2010 (Hansen et al., 2005). The French Agency for Environment and Energy Management (ADEME) estimates that the 2010 objective would require industrial rapeseed plantation in EU to increase from currently 7.4 million acres to 20 million acres (USDA, 2003). A longer term projection is shown in Table 14.2 for the shares of various alternative fuels in transportation, under the optimistic development scenario of the EC (Demirbas, 2008a, 2008c). The EU accounted for nearly 89% of all biodiesel production worldwide in 2005. By 2010, the United States is expected to become the world's largest single biodiesel market, accounting for roughly 18% of world biodiesel consumption, followed by Germany.

In Germany, the current aggressive program for development of the biodiesel industry is not a special exemption from EU law, but rather is based on a loophole in the law. The transportation fuel tax in Germany is based on petroleum fuel. Because biofuel is not a petroleum fuel, it can be used without being taxed. Unlike France and Italy, where biodiesel is blended with petroleum diesel, biodiesel sold in Germany is pure (100%) vegetable oil methyl ester. Hence, when taxed diesel prices are high and vegetable oil prices are low, biodiesel becomes very profitable. Additionally, with no restrictions on the quantity of biodiesel that can be exempted from the tax, there has been a huge investment in biodiesel production capacity (USDA, 2003).

14.4 Chinese Biofuel Policy

The increasing thirst for energy to fuel its fast growing economy has made China keen to explore the potential of biofuel (Yang, Zhou, and Liu, 2009). China's biofuel development exemplifies the general global picture. Its biofuel industry is currently dominated by ethanol, produced mainly from corn, but other feedstocks such as wheat and cassava are used in limited quantities as well. Most of the production and use of biodiesel is in the experimental stage. In 2002, China started the ethanol program by mandating the blending into gasoline in several big cities. Later, a compulsory use of a 10% blend was introduced in several provinces in October 2004, and extended to 27 other cities in 2006. To support domestic ethanol production, the government provides CNY 1.5 billion ($188 million) per year in financial subsidies to ethanol producers. In 2006, the production subsidy was CNY 1,373 (or $172) per ton. In addition, the value-added taxes are refunded for ethanol production, and fuel is exempt from the 5% consumption tax. Finally, the government covered any loss due to processing, transportation, or sale of the ethanol-blended fuel. Ethanol production was put forward in trial in 2002, and commercial supply became available in 2004. By 2007, the total ethanol production had reached 1.73 million metric tons or 0.58 billion gallons (Cheng, 2007).

The governments of developing countries, such as China, have well-intentioned biofuel programs to deal with energy security and environmental problems, and to increase farm income. However, securing feedstock is a crucial problem because these feedstocks are used as food in many countries. Increasing first-generation biofuel consumption can exacerbate the problem. For example, increasing food prices, due to competition, could lead to social unrest. Hence, the future focus will be on the second-generation biofuels that do not compete with the food supply. As various countries develop second-generation biofuel, international cooperation is needed to maximize such development in Asia and other developing regions (Koizumi and Ohga, 2007).

14.5 Indian Biofuel Policy

The Indian government is currently implementing an ethanol-blending program and considering a biodiesel-blending initiative. Biofuels are likely to play a significant role in India with increasing liquid fuel demand in the transportation sector due to rising population and its standard of living. On September 12, 2008, the Indian government announced National Biofuel Policy, which aims to meet 20% of India's diesel demand with fuel derived from plants. That will mean setting aside 35 million acres of land, from the present fuel-yielding plantation of less than 1.2 million acres. Salient features of the policy are

- An indicative target of 20% by 2017 for the blending of biofuel (ethanol and biodiesel).
- Biodiesel production will be taken up from nonedible oil seeds in waste/degraded/marginal lands.
- The focus would be on indigenous production of biodiesel feedstock, and import of vegetable oil for biodiesel production would not be permitted.
- Biodiesel plantations on community/government/forest waste lands would be encouraged, whereas plantation in fertile irrigated lands would not be encouraged.
- Minimum support price (MSP) with the provision of periodic revision for biodiesel oil seeds would be announced to provide fair price to the growers.

The details about the MSP mechanism, enshrined in the National Biofuel Policy, would be worked out carefully subsequently and considered by the Biofuel Steering Committee.

- Minimum purchase price (MPP) for the purchase of ethanol by the Oil Marketing Companies would be based on the actual cost of production and import price of ethanol biofuel. In the case of biodiesel, the MPP should be linked to the prevailing retail diesel price.
- The National Biofuel Policy envisages that biofuel, namely biodiesel and ethanol, may be brought under the ambit of "Declared Goods" by the government to ensure unrestricted movement of biofuel within and outside the States.

Due to these strategies, the rising population, and the growing energy demand from the transport sector, biofuel can be assured a significant market in India. The main policy drivers are historic, functional, economic, environmental, moral, and political. Ethanol production is centered on the use of molasses from the sugarcane industry. And the biodiesel production in India centers mainly around the cultivation and processing of jatropha plant seeds, which are very rich in oil (40 wt%). The estimated oil yield from jatropha plantations is about 3,000 kg/acre compared with 926 kg/acre per for soybeans in the United States and 2,470 kg/acre for rapeseed in Europe (Gopinathan and Sudhakaran, 2009).

14.6 The United States Biofuel Policy

Driven mainly by concerns over energy security and greenhouse gas emissions, the U.S. biofuel industry is undergoing rapid growth and transformation. National-level and state policies are being developed and implemented to promote much greater use of biofuels. Some of the states (e.g., California) have taken a strong position with respect to biofuel (Hoekman, 2009). The initial biofuel efforts have mainly been regarding corn ethanol. President Bush spoke in his January 31, 2006, State of the Union address about producing biofuel by 2012 using "woodchips, stalks, and switchgrass" as the source of cellulosic biomass (Demirbas, 2008b). Hence, significant research and development efforts are underway to develop second-generation biofuels produced from a variety of biomass and a wide range of conversion technologies.

The Energy Policy Act of 2005 was a significant step for the development of renewable energy. This $14 billion national energy plan contained numerous provisions related to energy efficiency and conservation, modernization of energy infrastructure, and promotion of both traditional energy sources and renewable alternatives. There were several provisions to spur the development of biofuels, including enhanced collaboration among government, industry, and academic institutions to develop advanced technologies for production of biofuels. The production incentives were placed to ensure the production of 1 billion gallons/year cellulosic biofuel from nonedible plant materials by 2015 (Hoekman, 2009).

In 2007, President Bush announced a much more aggressive program to reduce gasoline consumption and further increase the use of biofuel (Bush, 2007). The

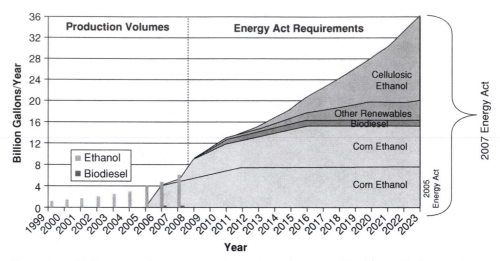

Figure 14.1. U.S. renewable fuel production and requirements (Hoekman, 2009; reproduced with permission from Elsevier).

so-called 20-in-10 Plan calls for a 20% reduction in the consumption of conventional fuels by the year 2017. This reduction will result from a combination of both increases in fuel efficiency and increases in use of renewable fuels. In 2007, the United States enacted the Energy Independence and Security Act of 2007 (U.S. Congress, 2007), which put in place much more ambitious goals for renewable fuels. The bill requires the use of 36 billion gallons/year of biofuels by 2022; and 21 out of 36 billion gallons/year biofuel must be advanced biofuel, meaning second-generation biofuel (Figure 14.1). The renewable fuel standard includes a provision for 1 billion gallons/year of biodiesel by 2012. Also, for the first time, the 2007 Energy Act includes the concept of a low-carbon fuel standard, requiring renewable fuels to have at least a 20% reduction in carbon intensity over the fuels' life cycle (Hoekman, 2009).

14.7 Global Biofuel Projections

Globally, renewable energy resources are more evenly distributed than fossil or nuclear resources, and the energy flows from the renewable resources are more than three orders of magnitude higher than the current use. Current energy system is unsustainable because of equity issues as well as environmental, economic, and geopolitical concerns that have implications far into the future (UNDP, 2000).

Biofuel is primarily agriculture energy, and its development can contribute to the economic growth of developing countries, as in these countries economy is based on agricultural production and most people live in rural areas. The transition from a fossil fuel to biofuel economy should be integrated with the community development programs, which can improve the socioeconomic development of the country. Various scenarios have shown that the introduction of biofuel economy can provide growth in the gross domestic production and per capita affluence and reduce reliance on the petroleum imports (Demirbas, 2006c). However, the availability of the resources is an important factor for the use of biofuel in the electricity, heat, or liquid fuel markets.

Demand for energy is continuously increasing due to the rapid outgrowth of population and urbanization. As the major conventional energy resources (i.e., coal, petroleum, and natural gas) are depleting, biomass can be considered as one of the promising environment-friendly renewable energy options. In the most biomass-intensive scenario, modernized biomass energy will contribute about 50% of total energy demand in developing countries by 2050 (IPCC, 1997). This scenario includes 1 billion acres of biomass energy plantations globally by 2050, with 75% of this area established in developing countries (Kartha and Larson, 2000).

According to the International Energy Agency, scenarios developed for the United States and Europe indicate that near-term targets of up to 6% displacement of petroleum fuels with biofuel appear feasible using conventional biofuel, given available cropland. A 5% displacement of gasoline in the EU requires about 5% of available cropland to produce ethanol, whereas in the United States, the requirement increases to 8%. A 5% displacement of diesel requires 13% of U.S. cropland and 15% of EU cropland (IEA, 2004).

Many farmers who grow oilseeds use a biodiesel blend in their farm equipment to foster production of biodiesel and raise public awareness. Hence, it is sometimes easier to find biodiesel in rural areas than in cities. To fully evaluate the policy options, additional factors must be taken into account, such as the fuel equivalent of the energy required for biodiesel processing, the yield of biodiesel from raw oil, the return on cultivating food, and the relative cost of biodiesel versus petroleum diesel. Given the history of petroleum politics, it is imperative that today's policy decisions ensure a free market for biodiesel, in which producers of all sizes compete in this industry, without farm subsidies, regulations, and other interventions skewing the market place (Eckhardt, 2005; Wardle, 2003; USDA, 2006).

14.8 Summary

A number of countries have implemented policies to foster growth of biofuels to partly replace the petroleum fuels or to meet future energy needs. Policy drivers for renewable biofuel have attracted particularly high levels of assistance in some countries, given their promise of benefits in agricultural production, trade balances, and rural development and economic opportunities. First-generation biofuels required government subsidy and mandatory use regulations. However, to sustain the long-term use of biofuel, now focus should be on implementing policies for second-generation biofuels that are derived from biomass noncompeting with the food supply. With judicious policies and their implementations, governments can meet their targets for decrease in CO_2 emissions and air pollution, petroleum import reduction, fuel security, general economic growth, and rural development.

References

Abbaslou, M., Reza, M., Soltan Mohammadzadeh, J. S., Dalai, A. K. 2009. Review on Fischer-Tropsch synthesis in supercritical media. *Fuel Process Technol* 90(7–8):849–856.

ACE. 2009. Petroleum. Available from: http://www.speedace.info/petroleum.htm. Accessed March 11, 2009.

Achten, W. M. J., Verchot, L., Franken, Y. J., et al. 2008. Jatropha bio-diesel production and use. *Biomass Bioenerg* 32(12):1063–1084.

Adjaye, J. D., Sharma, R. K., Bakhshi, N. N. 1992. Characterization and stability analysis of wood-derived bio-oil. *Fuel Process Technol* 31(3):241–256.

Agbogbo, F. K., Coward-Kelly, G. 2008. Cellulosic ethanol production using the naturally occurring xylose-fermenting yeast, Pichia stipitis. *Biotechnol Lett* 30(9):1515–1524.

Agbogbo, F. K., Haagensen, F. D., Milam, D., Wenger, K. S. 2008. Fermentation of acid-pretreated corn stover to ethanol without detoxification using Pichia stipitis. *Appl Biochem Biotechnol* 145(1–3):53–58.

Agranet. 2009. F.O. Licht's World Ethanol & Biofuels Report. Available from: http://www .agra-net.com/portal2/home.jsp?template=productpage&pubid=ag072.

Aitani, A. M. 2004. Oil refining and products. In: Cleveland, C. J., editor. Encyclopedia of Energy. New York, NY, Elsevier, pp. 15–729.

Antal, M. J., Hofmann, L., Moreira, J. R., Brown, C. T., Steenblick, R. 1981. Design and operation of a solar fired biomass flash pyrolysis reactor. IGT's Symposium: Energy from Biomass and Wastes V, Lake Buena Vista, Florida.

Appell, H. R. 1977. The production of oil from wood waste. In: Anderson, L., Tilman, D. A., editors. Fuels from Waste. New York, NY, Academic Press, pp. 121–140.

Archer, D., Wang, P. 1990. The dielectric constant of water and Debye–Huckel limiting law slopes. *J. Phys. Chem. Ref. Data* 19:371–411.

[ASTM] American Society for Testing and Materials. 2003. ASTM G173–03. Available from: http://rredc.nrel.gov/solar/spectra/am1.5/. Retrieved August 21, 2009.

ASTM. 2009. ASTM D6751-09 standard specification for biodiesel fuel blend stock (B100) for middle distillate fuels. Available from: http://www.astm.org/Standards/D6751.htm.

Authier, O., Ferrer, M., Mauviel, G., Khalfi, A.-E., Lede, J. 2009. Wood fast pyrolysis: Comparison of lagrangian and eulerian modeling approaches with experimental measurements. *Ind Eng Chem Res* 48(10):4796–4809.

[AWEA] American Wind Energy Association. 2005. The economics of wind energy. Available from: http://www.awea.org/pubs/factsheets/EconomicsOfWind-Feb2005.pdf and http://www.awea.org/faq/cost.html.

Babu, B. V., Chaurasia, A. S. 2003. Modeling for pyrolysis of solid particle: Kinetics and heat transfer effects. *Energ Convers Manage* 44(14):2251–2275.

Backman, R., Frederick, W. J., Hupa, M. 1993. Basic studies on black-liquor pyrolysis and char gasification. *Bioresource Technol* 46(1–2):153–158.

Bai, F. W., Anderson, W. A., Moo-Young, M. 2008. Ethanol fermentation technologies from sugar and starch feedstocks. *Biotechnol Adv* 26(1):89–105.

Bala, B. K. 2005. Studies on biodiesels from transformation of vegetable oils for diesel engines. *Energ Edu Sci Technol* 15(1–2):1–45.

Balat, M. 2007. An overview of biofuels and policies in the European Union countries. *Energ Sources* 2(2):167–181.

Bandura, A. V., Lvov, S. N. 2006. The ionization constant of water over wide ranges of temperature and density. *J Phys Chem Ref Data* 35(1):15–30.

Bauen, A., Woods, J., Hailes, R. 2004. BIOPOWERSWITCH! A biomass blueprint to meet 15% of OECD electricity demand by 2020. Prepared for WWF International. Available from: http://assets.panda.org/downloads/biomassreportfinal.pdf.

Beaumont, O. 1985. Flash pyrolysis products from beech wood. *Wood Fiber Sci* 17:228–239.

Bech, N., Larsen, M. B., Jensen, P. A., Dam-Johansen, K. 2009. Modeling solid-convective flash pyrolysis of straw and wood in the pyrolysis centrifuge reactor. *Biomass Bioenerg* 33(6–7):999–1011.

Becker, J., Boles, E. 2003. A modified Saccharomyces cerevisiae strain that consumes l-arabinose and produces ethanol. *Appl Environ Microbiol* 69(7):4144–4150.

Berl, E. 1944. Production of oil from plant material. *Science* 99(2573):309–312.

Bhattacharya, S. C., Abdul Salam, P. 2002. Low greenhouse gas biomass options for cooking in the developing countries. *Biomass Bioenerg* 22:305–317.

Biox Corporation. 2009. Production process. Available from: http://www.bioxcorp.com/production_process.php.

Blanco, M. I. 2009. The economics of wind energy. *Renew Sust Energ Rev* 13(6–7):1372–1382.

Bobleter, O. 1994. Hydrothermal degradation of polymers derived from plants. *Prog Polym Sci* 19:797–841.

Bolinger, M., Wiser, R. 2006. A comparative analysis of business structures suitable for farmer-owned wind power projects in the United States. *Energ Policy* 34(14):1750–1761.

Boocock, D. G. B. 2001. Single-phase process for production of fatty acid methyl esters as biofuels from mixtures of triglycerides and fatty acids. PCT Int. Appl., 26 pp., CODEN: PIXXD2 WO 2001012581 A1 20010222.

Boocock, D. G. B., Konar, S. K., Mao, V., Sidi, H. 1996. Fast one-phase oil-rich processes for the preparation of vegetable oil methyl esters. *Biomass Bioenerg* 11(1):43–50.

Boutin, O., Ferrer, M., Lédé, J. 1998. Radiant flash pyrolysis of cellulose: Evidence for the formation of short life time intermediate liquid species. *J Anal Appl Pyrol* 47(1):13–31.

BP. 2009. Statistical Review of World Energy. Available from: http://bp.com/statisticalreview. Accessed June 2009.

Branca, C., Di Blasi, C. 2006. Multistep mechanism for the devolatilization of biomass fast pyrolysis oils. *Ind Eng Chem Res* 45(17):5891–5899.

Bridgwater, A. V. 2003. Renewable fuels and chemicals by thermal processing of biomass. *Chem Eng J* 91(2–3):87–102.

Bridgwater, A. V., Meier, D., Radlein, D. 1999. An overview of fast pyrolysis of biomass. *Org Geochem* 30(12):1479–1493.

Brown, R. C. 2003. Biorenewable Resources: Engineering New Products from Agriculture. Ames, IA, Iowa State University Press.

Brown, R. C., Holmgren, J. 2009. Fast pyrolysis and bio-oil upgrading. Available from: http://www.ars.usda.gov/sp2UserFiles/Program/307/biomasstoDiesel/RobertBrown&Jennifer Holmgrenpresentationslides.pdf. Accessed June 2009.

Brunner, G. 2009. Near critical and supercritical water. Part I. Hydrolytic and hydrothermal processes. *J Supercrit Fluid* 47:373–381.

Bukur, D. B., Nowicki, Z., Manne, R. K., Lang, X. 1995. Activation studies with a precipitated iron catalysts for Fischer-Tropsch synthesis. II. Reaction studies. *J Catal* 155(2):366–375.

Bush, G. W. 2007. State of the Union Address. Available from: http://georgewbush-whitehouse.archives.gov/news/releases/2007/01/20070123-2.html Washington, DC.

Byrd, A. J., Pant, K. K., Gupta, R. B. (2007). Hydrogen production from ethanol by reforming in supercritical water using Ru/Al$_2$O$_3$ catalyst. *Energ Fuel* 21(6):3541–3547.

Byrd, A. J., Pant, K. K., Gupta, R. B. (2008). Hydrogen production from glycerol by reforming in supercritical water over Ru/Al$_2$O$_3$ catalyst. *Fuel* 87(13–14):2956–2960.

[CDIAC] Carbon Dioxide Information Analysis Center. 2009. Oak Ridge, TN, Ridge National Laboratory. Available from: http://cdiac.ornl.gov.

Chandel, A. K., Narasu, M. L., Chandrasekhar, G., Manikyam, A., Rao, L. V. 2009. Use of Saccharum spontaneum (wild sugarcane) as biomaterial for cell immobilization and modulated ethanol production by thermotolerant Saccharomyces cerevisiae VS3. *Bioresour Technol* 100(8):2404–2410.

Chandra, R. P., Bura, R., Mabee, W. E., Berlin, A., Pan, X., Saddler, J. N. 2007. Substrate pretreatment: The key to effective enzymatic hydrolysis of lignocellulosics? *Adv Biochem Eng Biotechnol* 108: 67–93.

Chen, Z., Blaabjerg, F. 2009. Wind farm – A power source in future power systems. *Renew Sust Energ Rev* 13(6–7):1288–1300.

Cheng, Q. 2007. The impact of biofuel development on grain market. *Heilongjiang Grain* 4:26–29.

Chongkhong, S., Tongurai, C., Chetpattananondh, P. 2009. Continuous esterification for biodiesel production from palm fatty acid distillate using economical process. *Renew Energ* 34(4):1059–1063.

Chornet, E., Overend, R. P. 1985. Biomass liquefaction: An overview. In: Overend, R. P., Milne, T. A., Mudge, L. K., editors. Fundamentals of Thermochemical Biomass Conversion. New York, NY, Elsevier Applied Science, pp. 967–1002.

Chum, H. L., Black, S. K. 1990. Process for fractionating fast-pyrolysis oils, and products derived therefrom. U.S. Patent 4942269.

CleanTechnica. 2008. Available from: http://cleantechnica.com/files/2008/03/wood1.jpg.

Colares, J. F. 2008. A brief history of Brazilian biofuel legislation. *Syracuse J Law Commerce* 35:101–116.

[COSO] Crude Oil Supply Outlook. 2007. Report to the Energy Watch Group EWG-Series No 3/2007, October. Available from: http://www.energywatchgroup.org/fileadmin/global/pdf/EWG_Oilreport_10–2007.pdf.

Cuff, D. J., Young, W. J. 1980. The United States Energy Atlas. New York, NY, Free Press/Macmillan.

Czernik, S., Bridgwater, A. V. 2004. Overview of applications of biomass fast pyrolysis oil. *Energ Fuel* 18(2):590–598.

Dale, B. E., Leong, C. K., Pham, T. K., Esquivel, V. M., Rios, I., Latimer, V. M. 1996. Hydrolysis of lignocellulosics at low enzyme levels: Application of the AFEX process. *Bioresource Technol* 56(1):111–116.

Davis, B. H. 2002. Overview of reactors for liquid phase Fischer–Tropsch synthesis. *Catal Today* 71:249–300.

Davis, H., Figueroa, C., Schaleger, L. 1982. Hydrogen or carbon monoxide in the liquefaction of biomass. *Adv Hydrogen Energ* 3(2):849–862.

Davis, K. S. 2001. Corn Milling, Processing and Generation of Co-products. Minnesota Nutrition Conference, Minnesota Corn Growers Association Technical Symposium, September 11, 2001.

Davis, S. C., Anderson-Teixeira, K. J., DeLucia, E. H. 2009. Life-cycle analysis and the ecology of biofuels. *Trends Plant Sci* 14(3):140–146.

De La Torre Ugarte, D. G., English, B. C., Jensen, K. 2007. Sixty billion gallons by 2030: Economic and agricultural impacts of ethanol and biodiesel expansion. *Am J Agr Econ* 89(5):1290–1295.

De Wit, M., Faaij, A. 2009. European biomass resource potential and costs. *Biomass Bioenerg*, Available online August 21, 2009, doi:10.1016/j.biombioe.2009.07.011.

Demirbas, A., Gullu, D., Caglar, A., Akdeniz, F. 1997. Determination of calorific values of fuel from lignocellulosics. *Energ Source* 19: 765–770.

Demirbas, A. 2000a. Mechanisms of liquefaction and pyrolysis reactions of biomass. *Energ Convers Manage* 41:633–646.

Demirbas, A. 2000b. Recent advances in biomass conversion technologies. *Energ Edu Sci Technol* 6:19–41.

Demirbas, A. 2000c. Biomass resources for energy and chemical industry. *Energ Edu Sci Technol* 5:21–45.

Demirbas, A. 2002a. Biodiesel from vegetable oils via transesterification in supercritical methanol. *Energ Convers Manage* 43:2349–56.

Demirbas, A. 2002b. Pyrolysis and steam gasification processes of black liquor. *Energ Convers Manage* 43:877–884.

Demirbas, A. 2003. Biodiesel fuels from vegetable oils via catalytic and non-catalytic supercritical alcohol transesterifications and other methods: A survey. *Energ Convers Manage* 44:2093–2109.

Demirbas, A. 2004a. Ethanol from cellulosic biomass resources. *Int J Green Energ* 1:79–87.

Demirbas, A. 2004b. Combustion characteristics of different biomass fuels. *Prog Energ Combust* 30(2):219–230.

Demirbas, A. 2005a. Pyrolysis of ground beech wood in irregular heating rate conditions. *J Anal Appl Pyrolysis* 73:39–43.

Demirbas, A. 2005b. Bioethanol from cellulosic materials: A renewable motor fuel from biomass. *Energ Source* 27:327–337.

Demirbas, A. 2006a. Electrical power production facilities from green energy sources. *Energ Sources* 1: 291–301.

Demirbas, A. 2006b. Energy priorities and new energy strategies. *Energ Edu Sci Technol* 16:53–109.

Demirbas, A. 2006c. Global biofuel strategies. *Energ Edu Sci Technol* 17:27–63.

Demirbas, A. 2007a. Importance of biodiesel as transportation fuel. *Energ Policy* 35: 4661–4670.

Demirbas, A. 2007b. Progress and recent trends in biofuels. *Prog Energ Combust* 33:1–18.

Demirbas, A. 2007c. The influence of temperature on the yields of compounds existing in bio-oils obtained from biomass samples via pyrolysis. *Fuel Proc Technol* 88:591–597.

Demirbas, A. 2008a. Biodiesel: A Realistic Fuel Alternative for Diesel Engines. London, Springer.

Demirbas, A. 2008b. Economic and environmental impacts of the liquid biofuels. *Energ Edu Sci Technol* 22:37–58.

Demirbas, A. 2008c. Biofuel sources, biofuel policy, biofuel economy and global biofuel projections. *Energ Convers Manage* 49:2106–2116.

Demirbas, A. H. 2009. Inexpensive oil and fats feedstocks for production of biodiesel. *Energ Edu Sci Technol A* 23(1–2):1–13.

Demirbas, A., Arin, G. 2002. An overview of biomass pyrolysis. *Energ Source* 24:471–482.

Demirbas, M. F. 2007. Electricity production using solar energy. *Energ Source A* 29:563–569.

Demirbas, M. F., Balat, M. 2006. Recent advances on the production and utilization trends of bio-fuels: A global perspective. *Energ Convers Manage* 47(15–16):2371–2381.

Deng, L., Yan, Z., Fu, Y., Guo, Q. X. 2009. Green solvent for flash pyrolysis oil separation. *Energ Fuel* 23(6):3337–3338.

Diaz, E., Mohedano, A. F., Calvo, L., Gilarranz, M. A., Casas, J. A., Rodriguez, J. J. 2007. Hydrogenation of phenol in aqueous phase with palladium on activated carbon catalysts. *Chem Eng J* 131(1–3):65–71.

Diebold, J. P. 2000. A review of the chemical and physical mechanisms of the storage stability of fast pyrolysis bio-oil. Available from: http://www.nrel.gov/docs/fy00osti/27613.pdf.

Diebold, J. P., Czernik, S. 1997. Additives to lower and stabilize the viscosity of pyrolysis oils during storage. *Energ Fuel* 11(5):1081–1091.

[DNR] Department of Natural Resources, Louisiana. 2009. Basic ethanol production. Available from: http://dnr.louisiana.gov/sec/execdiv/TECHASMT/alternative_fuels/ethanol/fuel_alcohol_1987/015.htm. Accessed August 2009.

Dry, M. E. 2002. The Fischer–Tropsch process: 1950–2000. *Catal Today* 71:227–241.

Dufey, A. 2006. Biofuels production, trade and sustainable development: Emerging issues. Environmental Economics Programme, Sustainable Markets Discussion Paper No. 2. London, International Institute for Environment and Development (IIED), September.

Eckhardt, A., 2009. Freedom Fuel: How and Why Biodiesel Policy Should Reflect Freedom. Available from: http://www.cascadepolicy.org/2005/10/14/freedom-fuel-how-and-why-biodiesel-policy-should-reflect-freedom/. Accessed Dec 3, 2009.

[EC] European Commission. 2003. Renewable energies: A European policy. Promoting Biofuel in Europe. European Commission, Directorate-General for Energy and Transport, B-1049, Bruxelles, Belgium.

Edinger, R., Kaul, S. 2000. Humankind's detour toward sustainability: Past, present, and future of renewable energies and electric power generation. *Renew Sust Energ Rev* 4:295–313.

[EIA] Energy Information Agency. 2009a. Available from: http://www.eia.doe.gov/.

EIA. 2009b. Energy KIDS: Renewable hydropower. Available from: http://www.eia.doe.gov/kids/energyfacts/sources/renewable/water.html.

Elbashir, N. O., Roberts, C. B. 2004. Selective control of hydrocarbon product distribution in supercritical phase Fischer–Tropsch synthesis. *ACS Div Petrol Chem Prepr* 49:422–425.

Elliott, D. C. 2007. Historical developments in hydroprocessing bio-oils. *Energ Fuel* 21(3):1792–1815.

Elliott, D. C., Schiefelbein, G. F. 1989. Liquid hydrocarbon fuels from biomass. *Amer Chem Soc Div Fuel Chem Preprints* 34(4):1160–1166.

Encinar, J. M., Gonzalez, J. F., Rodríguez, J. J., Tejedor, A. 2002. Biodiesel fuels from vegetable oils: Transesterification of Cynara cardunculus L. oils with ethanol. *Energ Fuel* 16:443–450.

[EREC] European Renewable Energy Council. 2006. Renewable Energy Scenario by 2040. Brussels, EREC Statistics.

Escobar, J. C., Lora, E. S., Venturini, O. J., Yanez, E. E., Castillo, E. F., Almazan, O. 2009. Biofuels: Environment, technology and food security. *Renew Sust Energ Rev* 13(6–7):1275–1287.

[ESRL] Earth System Research Laboratory. 2009. Teacher resources: Carbon cycle toolkit. Available from: http://www.esrl.noaa.gov/gmd/education/carbon_toolkit/basics.html.

[EWEA] European Wind Energy Association. 2005. Report: Large scale integration of wind energy in the European power supply: Analysis, issues and recommendations. Available from http://www.ewea.org/fileadmin/ewea_documents/documents/publications/grid/051215_Grid_report.pdf.

Fang, Z., Minowa, T., Smith, R. L., Jr., Ogi, T., Kozinski, J. A. 2004. Liquefaction and Gasification of Cellulose with Na2CO3 and Ni in Subcritical Water at 350 C. *Ind. Eng. Chem. Res.* 43:2454–2463.

Fang, Z., Sato, T., Smith, R. L., Jr., Inomata, H., Arai, K., Kozinski, J. A. 2008. Reaction chemistry and phase behaviour of lignin in high-temperature and super critical water. *Bioresource Technol* 99:3424–3430.

Feng, W., van der Kooi, H. J., de Swaan Arons, Jakob. 2004a. Phase equilibria for biomass conversion processes in subcritical and supercritical water. *Chem Eng J* 98:105–113.

Feng, W., van der Kooi, H. J., de Swaan Arons, J. 2004b. Biomass conversions in subcritical and supercritical water: driving force, phase equilibria, and thermodynamic analysis. *Chem Eng Process* 43:1459–1467.

Foster, B. L., Dale, B. E., Doran-Peterson, J. B. 2001. Enzymatic hydrolysis of ammonia-treated sugar beet pulp. *Appl Biochem Biotechnol* 91(3):269–282.

Fratzl, P. 2003. Cellulose and collagen: from fibres to tissues. *Curr Opin Colloid In* 8(1):32–39.

Fridleifsson, I. B. 2001. Geothermal energy for the benefit of the people. *Renew Sust Energ Rev* 5:299–312.

Galbe, M., Sassner, P., Wingren, A., Zacchi, G. 2007. Process engineering economics of bioethanol production. *Adv Biochem Eng Biotechnol* 108:303–327.

Galbe, M., Zacchi, G. 2007. Pretreatment of lignocellulosic materials for efficient bioethanol production. *Adv Biochem Eng Biotechnol* 108:41–65.

Garcia-Perez, M., Chaala, A., Roy, C. 2002. Vacuum pyrolysis of sugarcane bagasse. *J Anal Appl Pyrol* 65:111–136.

Garg, H. P., Datta, G. 1998. Global status on renewable energy. In: Solar Energy Heating and Cooling Methods in Building. From the International Workshop of Iran University of Science and Technology, May 19–20.

Gavillan, R. M., Mattschei, P. K. 1980. Fractionation of oil obtained by pyrolysis of ligno-cellulosic materials to recover a phenolic fraction for use in making phenol-formaldehyde resins. U.S. Patent 4233465.

[GBEP] Global Bioenergy Partnership. 2007. A review of the current state of bioenergy development in G8+5 countries, New York, NY.

Gleick, P. H. 1999. The World's Water: The Biennial Report on Freshwater Resources. Oakland, CA, Pacific Institute for Studies in Development, Environment, and Security.

Goddard Institute for Space Studies. 2009. Datasets and images. Available from: http://data.giss.nasa.gov/gistemp/graphs. Accessed August 20, 2009.

Goldemberg, J. 2002. Brazilian Energy Initiative. World Summit on Sustainable Development, Johannesburg, South Africa.

Goldemberg, J., Coelho, S. T., Nastari, P. M., Lucon, O. 2004. Ethanol learning curve-the Brazilian experience. *Biomass Bioenerg* 26:301–304.

Goldemberg, J., Coelho, S. T., Guardabassi, P. 2008. The sustainability of ethanol production from sugarcane. *Energ Policy* 36(6):2086–2097.

Goldstein, I. S. 1981. Organic Chemicals from Biomass. Boca Raton, FL, CRC Press.

Gopinathan, M. C., Sudhakaran, R. 2009. Biofuels: Opportunities and challenges in India. *In Vitro Cell Dev Biol Plant* 45(3):350–371.

Goudriaan, F., Peferoen, D. G. R. 1990. Liquid fuels from biomass via a hydrothermal process. *Chem Eng Sci* 45(8):2729–2734.

Goyal, H. B., Seal, D., Saxena, R. C. 2007 (volume date 2008). Biofuels from thermochemical conversion of renewable resources: A review. *Renewable & Sustainable Energy Reviews* 12:504–517.

Graham, L. A., Belisle, S. L., Baas, C.-L. 2008. Emissions from light duty gasoline vehicles operating on low blend ethanol gasoline and E85. *Atmos Environ* 42: 4498–4516.

Grebner, D. L., Perez-Verdin G., Sun C., Munn I. A., Schultz, E. B., Mamey, T. G. 2008. Woody biomass feedstocks; A case study on availability, production costs, and implications for bioenergy conversion in Mississippi. In: Solomon, B., and Luzadis, V., editors. Renewable Energy From Forest Resources in the United States. New York, NY, Routledge, pp. 261–280.

Gronli, M. 1996. A theoretical and experimental study of the thermal degradation of biomass. PhD Thesis, University of Trondheim, Norway.

Guettel, R., Kunz, U., Turek, T. 2008. Reactors for Fischer-Tropsch synthesis. *Chem Eng Technol* 31(5):746–754.

Gui, M. M., Lee, K. T., Bhatia, S. 2008. Feasibility of edible oil vs. non-edible oil vs. waste edible oil as biodiesel feedstock. *Energy* 33(11):1646–1653.

Gunasekaran, P., Raj K. C. 1999. Ethanol fermentation technology: *Zymomonas mobilis*. *Curr Sci India* 77:56–68.

Gupta, R. B. 2008. Hydrogen Fuel: Production, Transport, and Storage. Boca Raton, FL, CRC Press.

Gutherz, J. M., Schiller, M. E. 1991. A passive solar heating system for the perimeter zone of office buildings. *Energy Source* 13(1):39–54.

[GWEC] Global Wind Energy Council. 2009. Available from: http://www.gwec.net/.

Haas, M. J., McAloon A. J., Yee, W. C., Foglia, T. A. 2006. A process model to estimate biodiesel production costs. *Bioresource technology* 97:671–8.

Hacisalihoglu, B., Demirbas, A. H., Hacisalihoglu, S. 2008. Hydrogen from gas hydrate and hydrogen sulfide in the Black Sea. *Energ Edu Sci Technol* 21:109–115.

Hall, D. O., Mynick, H. E., Williams, R. H. 1991. Carbon sequestration versus fossil fuel substitution: alternative roles for biomass in coping with greenhouse warming. In: White, J. C., editor. Global Climate Change: the Economic Costs of Mitigation and Adaptation. New York, NY, Elsevier Science, pp. 241–282.

Hansen, A. C., Zhang, Q., Lyne, P. W. L. 2005. Ethanol-diesel fuel blends: A review. *Bioresource Technol* 96(3):277–285.

Hartley, I. D., Wood, L. J. 2008. Hygroscopic properties of densified softwood pellets. *Biomass Bioenerg* 32(1):90–93.

Hashaikeh, R., Fang, Z., Butler, I. S., Hawari, J., Kozinski, J. A. 2007. Hydrothermal dissolution of willow in hot compressed water as a model for biomass conversion. *Fuel* 86:1614–1622.

Hashem, A., Akasha, R. A., Ghith, A., Hussein, D. A. 2007. Adsorbent based on agricultural wastes for heavy metal and dye removal: A review. *Energ Edu Sci Technol* 19(1–2):69–86.

Hawes, D., Feldman, D., Banu, D. 1993. Latent heat storage in building materials. *Energ Buildings* 20:77–86.

He, B. J., Zhang, Y., Yin, Y., Funk, T. L., Riskowski, G. L. 2001. Effects of alternative process gases on the thermochemical conversion process of swine manure. *T ASAE* 44(6):1873–1880.

Hiete, M., Berner, U., Richter, O. 2001. Calculation of global carbon dioxide emissions: Review of emission factors and a new approach taking fuel quality into consideration. *Global Biogeochem Cy* 15(1):169–181.

Hill, J., Nelson, E., Tilman, D., Polasky, S., Tiffany, D. 2006. Environmental, economic, and energetic costs and benefits of biodiesel and ethanol biofuels. *Proc Natl Acad Sci USA* 103(30):11206–11210.

Hoekman, S. K. 2009. Biofuel in the US: Challenges and opportunities. *Renew Energ* 34:14–22.

Holmgren, J., Marinangeli, R., Nair, P., Elliott, D., Bain, R. 2008. Consider upgrading pyrolysis oils into renewable fuels. *Hydrocarb Process* 87(9):95–96, 98, 100, 103.

Hopkins, M. W., Antal, M. J. 1984. Radiant flash pyrolysis of biomass using a Xenon falsh tube. *J Appl Polymer Sci* 29:2163–2175.

Huang, C.-F., Lin, T.-H., Guo, G.-L., Hwang, W.-S. 2009. Enhanced ethanol production by fermentation of rice straw hydrolysate withgout detoxification using a newly adapted strain of Pichia stipitis. *Bioresource Technol* 100(17):3914–3920.

Huang, W. C., Ramey, D. E., Yang, S. T. 2004. Continuous production of butanol by Clostridium acetobutylicum immobilized in a fibrous bed bioreactor. *Appl Biochem Biotechnol A* 113–116:887–898.

Huang, X., Roberts, C. B. 2003. Selective Fischer–Tropsch synthesis over an Al_2O_3 supported cobalt catalyst in supercritical hexane. *Fuel Process Technol* 83(1–3):81–99.

[IEA] International Energy Agency. 2004. Biofuel for transport: An international perspective. Available from: http://www.iea.org/textbase/nppdf/free/2004/biofuels2004.pdf

IEA. 2007. Key world energy statistics, Paris. Available from: http://www.iea.org/Textbase/nppdf/free/2007/key_stats_2007.pdf.

IGP. 2009. Available from: http://www.industrialgasplants.com/gifs/floating-gas.jpg. Accessed April 2009.

[INL, DOE] Idaho National Laboratory, Department of Energy. 2009. Available from: https://inlportal.inl.gov/portal/server.pt?open=512&objID=422&parentname=Community Page&parentid=14&mode=2. Accessed August 2009.

[IPCC] Intergovernmental Panel on Climate Change. 1996. Climate Change 1995: The Science of Climate Change, Contribution of Working Group 1 to the Second Assessment Report of the IPCC, UNEP and WMO. Cambridge, Cambridge University Press.

IPCC. 1997. Greenhouse Gas Inventory Reference Manual: Revised 1996 IPCC Guidelines for National Greenhouse Gas Inventories, Report, 3:1.53, Paris. Available from: http://www.ipcc-nggip.iges.or.jp/public/2006gl/index.html.

IPCC. 2005. Carbon Dioxide Capture and Storage. IPCC Special Report prepared by Working Group III of the Intergovernmental Panel on Climate Change [Metz, B., O. Davidson, H.C. de Coninck, M. Loos and L.A. Meyer (eds.)], Cambridge University Press, Cambridge, United kingdom and New York, NY, USA, 442 pp.

Jacobson, D. L. 2007. PEM fuel cells. Available from: http://physics.nist.gov/MajResFac/NIF/pemFuelCells.html. Accessed September 7, 2007.

Janse, A. M. C., de Jong, X. A., Prins, W., van Swaaij, W. P. M. 1999. Heat transfer coefficients in the rotating cone reactor. *Powder Technol* 106(3):168–175.

Jansen, J. C. 2003. Policy support for renewable energy in the European Union. A review of the regulatory framework and suggestions for adjustment. Available from: http://www.ecn.nl/docs/library/report/2003/c03113.pdf.

Jean-Baptiste, P., Ducroux, R. 2003. Energy policy and climate change. *Energ Policy* 31:155–166.

Jean-Marie, A., Griboval-Constant, A., Khodakov, A. Y., Diehl, F. 2009. Cobalt supported on alumina and silica-doped alumina: Catalyst structure and catalytic performance in Fischer-Tropsch synthesis. *CR Acad Sci II C* 12(6–7):660–667.

Jeffries, T. W., Jin, Y. S. 2004. Metabolic engineering for improved fermentation of pentoses by yeasts. *Appl Microbiol Biotechnol* 63:495–509.

Jin, F. M., Zhou, Z. Y., Takehiko, M. 2005. Controlling hydrothermal reaction pathways to improve acetic acid production from carbohydrate biomass. *Environ Sci Technol* 39(6):1893–1902.

Jin, Y., Datye, A. K. 2000. Phase transformations in iron Fischer–Tropsch catalysts during temperature-programmed reduction. *J Catal* 196:8–17.

Jørgensen, H., Kristensen, J. B., Felby, C. 2007. Enzymatic conversion of lignocellulose into fermentable sugars: Challenges and opportunities. *Biofuels Bioprod Bioref* 1:119–134.

Jothimurugesan, K., Goodwin, J. G., Santosh, S. K., Spivey, J. J. 2000. Development of Fe Fischer–Tropsch catalysts for slurry bubble column reactors. *Catal Today* 58:335–344.

Jun, K.-W., Roh, H.-S., Kim, K.-Su., Ryu, J.-S., Lee, K.-W. (2004). Catalytic investigation for Fischer-Tropsch synthesis from bio-mass derived syngas. *Appl Catal A-Gen* 259(2):221–226.

Kadiman, O. K. 2005. Crops: Beyond foods. In: Proceedings of the 1st International Conference of Crop Security, Malang, Indonesia, September 20–23.

Kalogirou, S. A. 2004. Solar thermal collectors and applications. *Prog Energ Combust* 30:231–295.

Karagoez, S., Bhaskar, T., Muto, A., Sakata, Y., Oshiki, T., Kishimoto, T. 2005. Low-temperature catalytic hydrothermal treatment of wood biomass: analysis of liquid products. *Chem Eng J* 108(1–2):127–137.

Karhumaa, K., Wiedemann, B., Hahn-Hagerdal, B., Boles, E., Gorwa-Grauslund, M. F. 2006. Co-utilization of l-arabinose and d-xylose by laboratory and industrial Saccharomyces cerevisiae strains. *Microb Cell Fact* 10:5–18.

Karr, W. E., Holtzapple, M. T. 2000. Using lime pretreatment to facilitate the enzymatic hydrolysis of corn stover. *Biomass and Bioenergy* 18: 189–199.

Kartha, S., Larson, E. D. 2000. Bioenergy primer: Modernised biomass energy for sustainable development, Technical Report UN Sales Number E.00.III.B.6. New York, NY, United Nations Development Programme.

Kim, H., Choi, B. 2008. Effect of ethanol-diesel blend fuels on emission and particle size distribution in a common-rail direct injection engine with warm-up catalytic converter. *Renew Energ* 33: 2222–2228.

Kim, S., Dale, B. E. 2004. Global potential bioethanol production from wasted crops and crop residues. *Biomass Bioenerg* 26:361–375.

Kim Oanh, N. T., Upadhyay, N., Zhuang, Y.-H., et al. 2006. Particulate air pollution in six Asian cities: Spatial and temporal distributions, and associated sources. *Atmos Environ* 40(18):3367–3380.

Klinke, H. B., Thomsen, A. B., Ahring, B. K. 2004. Inhibition of ethanol-producing yeast and bacteria by degradation products produced during pre-treatment of biomass. *Appl Microbiol Biotechnol* 66:10–26.

Kodama, T. 2003. High-temperature solar chemistry for converting solar heat to chemical fuels. *Prog Energ Combust Sci* 29(6):567–597.

Kohl, A. L. 1986. Black liquor gasification. *Can J Chem Eng* 64:299–304.

Koizumi, T., Ohga, K. 2007. Biofuels Policies in Asian Countries: Impact of the Expanded Biofuels Programs on World Agricultural Markets. *Journal of Agricultural & Food Industrial Organization* 5:1–20.

Kong, L. Z., Li, G. M., Wang, H. 2008. Hydrothermal catalytic conversion of biomass for lactic acid production. *J Chem Technol Biotechnol* 83:383–388.

Kosugi, T., Pyong, S. P. 2003. Economic evaluation of solar thermal hybrid H_2O turbine. *Energy* 28:185–198.

Kruse, A., Dinjus, E. 2007a. Hot compressed water as reaction medium and reactant. 1. Properties and synthesis reactions. *J Supercrit Fluid* 39: 362–380.

Kruse, A., Dinjus, E. 2007b. Hot compressed water as reaction medium and reactant. 2. Degradation reactions. *J Supercrit Fluid* 41:361–379.

Kumar, P., Barrett, D. M., Delwiche, M. J., Stroeve, P. 2009. Methods for pretreatment of lignocellulosic biomass for efficient hydrolysis and biofuel production. *Ind Eng Chem Res* 48(8):3713–3729.

Kumar, S., Byrd, A., Gupta R. B. 2009. Sub- and super-critical water technology for biofuels: swtichgrass to ethanol, biocrude, and hydrogen. The 31st Symposium on Biotechnology for Fuels and Chemicals, San Francisco, May.

Kumar, S., Gupta, R. B. 2008. Hydrolysis of microcrystalline cellulose in sub- and supercritical water in a continuous flow reactor. *Ind Eng Chem Res* 47:9321–9329.

Kumar, S., Gupta, R. B. 2009. Biocrude production from switchgrass using subcritical water. *Energ Fuel*, 23:5151–5159.

Kusdiana, D., Saka, S. 2001. Kinetics of transesterification in rapeseed oil to biodiesel fuels as treated in supercritical methanol. *Fuel* 80:693–698.

Kusdiana, D., Saka, S. 2004a. Two-step preparation for catalyst-free biodiesel fuel production: Hydrolysis and methyl esterification. *Appl Biochem Biotechnol* 113–116:781–791.

Kusdiana, D., Saka, S. 2004b. Effects of water on biodiesel fuel production by supercritical methanol treatment. *Bioresource Technol* 91:289–295.

Kutz, M. (ed). 2007. Environmentally Conscious Alternative Energy Production. Hoboken, NJ, John Wiley & Sons.

Lachke, A. 2002. Biofuel from D-xylose – the second most abundant sugar. *Resonance* 7: 50–58.

Lal, R. 2005. World crop residues production and implications of its use as a biofuel. *Environ Int* 31:575–584.

Larson, E. D., Jin, H. 1999. Biomass conversion to Fischer-Tropsch liquids: Preliminary energy balances. In: Overend, R., Chornet, E., editors. Proceedings of the Fourth Biomass Conference of the Americas. Kidlington, Elsevier Science, vol. 1–2, pp. 843–854.

Laurent, E., Delmon, B. 1994. Influence of water in the deactivation of a sulfided NiMo/γ-A12O3 catalyst during hydrodexygenation. *J Catal* 146(1):284–291.

Laxman, R. S., Lachke, A. H. 2008. Bioethanol from lignocellulosic biomass. Part 1: Pretreatment of the substrates. In: Pandey, A., editor. Handbook of Plant-Based Biofuels. Boca Raton, FL, CRC Press, pp. 121–139.

Lédé, J., Pharabod, F. 1997. Chimie solaire dans le Monde et en France. *Entropie* 204: 47–55.

Lédé, J. 1998. Solar thermochemical conversion of biomass. *Solar Energ* 65(1):3–13.

Lédé, J., Berthelot, P., Villermaux, J., Rolin, A., François, H., Déglise, X. 1980. Pyrolyse flash de déchets ligno-cellulosiques en vue de leur valorisation par l'énergie solaire concentrée. *Rev Phys Appl* 15:545–552.

Lédé, J., Villermaux, J., Royère, C., Blouri, B., Flamant G. 1983. Utilisation de l'énergie solaire concentrée pour la pyrolyse du bois et des huiles lourdes du pétrole. *Entropie* 110: 57–69.

Lee, S. Y., Holder, G. D. 2001. Methane hydrates potential as a future energy source. *Fuel Process Technol* 71:181–186.

Leistritz, F. L., Hodur, N. M. 2008. Biofuels: A major rural economic development opportunity. *Biofuels Bioprod Bioref* 2(6):501–504.

Lin, S.-Y. 2009. Hydrogen production from coal. In: Gupta, R. B., editor. Hydrogen Fuel. Boca Raton, FL, CRC Press, pp. 103–125.

Low, S. A., Isserman, A. M. 2009. Ethanol and the local economy: Industry trends, location factors, economic impacts, and risks. *Econ Dev Q* 23(1):71–88.

Luijkx, G. C. A., Rantwijk, F. V., Bekkum, H. V. 1993. Hydrothermal formation of 1,2,4-benzenetriol from 5-hydroxymethyl-2-furaldehyde and D-fructose. *Carbohyd Res* 242:131–139.

Luo, Z. Y., Wang, S. R., Liao, S. R. 2004. Research on biomass fast pyrolysis for liquid fuel. *Biomass Bioenerg* 26(5):455–462.

Lurgi. 2009. Biodiesel. Available from: http://www.lurgi.com/website/fileadmin/user_upload/pdfs/02_Biodiesel-E.pdf.

Ma, F., Hanna, M. A. 1999. Biodiesel production: A review. *Bioresource Technol* 70:1–15.

Maine, F. W. 2006. Wood plastics composites workshop, June 15, 2006. Available from: http://www.seainnovation.com.

Mani, S., Tabil, L. G., Sokhansanj, S. 2004, Grinding performance and physical properties of wheat and barley straws, corn stover and switchgrass. *Biomass and Bioenergy* 27:339–352.

Martin, C., Galbe, M., Wahlbom, C. F., Hahn-Hagerdal, B., Jonsson, L. J. 2002. Ethanol production from enzymatic hydrolysates of sugarcane bagasse using recombinant xylose-utilising *Saccharomyces cerevisiae*. *Enzyme Microbial Technol* 31:274–282.

Masaru, W., Takafumi, S., Hiroshi, I. 2004. Chemical reactions of C1 compounds in near-critical and supercritical water. *Chem Rev* 104:5803–5821.

McKendry, P. 2002. Energy production from biomass (part 3): Gasification technologies. *Bioresource Technol* 83(1):55–63.

Mehtiev, S. F. 1986. Origin of petroleum. *Geol Balcan* 16(4):3–16.

Merino, S. T., Cherry, J. 2007. Progress and challenges in enzyme development for biomass utilization. *Adv Biochem Eng Biotechnol* 108:95–120.

Millet, M. A., Baker, A. J., Scatter, L. D. 1976. Physical and chemical pretreatment for enhancing cellulose saccharification. *Biotechnol Bioeng Symp* 6:125–153.

Minowa, T., Zhen, F., Ogi, T. 1998. Cellulose decomposition in hot-compressed water with alkali or nickel catalyst. *J Supercrit Fluid* 13:253–259.

Minowa, T., Zhen, F., Ogi, T. 1999. Liquefaction of cellulose in hot-compressed water using sodium carbonate: Production distribution at different reaction temperature. *J Chem Eng Jpn* 30(1):186–190.

Miyazawa, T., Funazukuri, T. 2005. Polysaccharide hydrolysis accelerated by adding carbon dioxide under hydrothermal conditions. *Biotechnol Prog* 21:1782–1785.

Mohan, D., Pittman, C. U., Jr., Steele, P. H. 2006. Pyrolysis of wood/biomass for bio-oil: A critical review. *Energ Fuel* 20:848–889.

Mok, W. S., Antal, M. J. 1992. Uncatalyzed solvolysis of whole biomass hemicellulose by hot compressed liquid water. *Ind Eng Chem Res* 31:1157–1161.

Monnet, F. 2003. An introduction to anaerobic digestion of organic wastes. A report by Remade Scotland. Available from: http://www.biogasmax.eu/media/introanaerobicdi gestion_073323000_1011_24042007.pdf.

Mulkins-Phillips, G. J., Stewart, J. E. 1974. Effect of environmental parameters on bacterial degradation of bunker C oil, crude oils, and hydrocarbons. *Appl Microbiol* 28:915–922.

Murphy, H., Niitsuma, H. 1999. Strategies for compensating for higher costs of geothermal electricity with environmental benefits. *Geothermics* 28:693–711.

Murray, J. P., Fletcher, E. A. 1994, Reaction of steam with cellulose in a fluidized bed using concentrated sunlight. *Energy* (Oxford, UK) 19:1083–1098.

Muthukumaran, P., Gupta, R. B. 2000. Sodium-carbonate-assisted supercritical water oxidation of chlorinated waste. *Ind Eng Chem Res* 39(12):4555–4563.

Nakamura, G., UC Cooperative Extension. 2004. Biomass thinning for fuel reduction and forest restoration: Issues and opportunities. Available from: http://ucce.ucdavis.edu/files/filelibrary/5098/16265.pdf. Accessed May 2009.

Nas, B., Berktay, A. 2007. Energy potential of biodiesel generated from waste cooking oil: an environmental approach. *Energ Sources* 2:63–71.

[NBII] National Biological Information Infrastructure, U.S. Geological Survey. 2009. Available from: http://images.nbii.gov/RFemmer/D_thumbnail/92 Sorghum 0 field of sorghum.jpg. Accessed August 2009.

Nelson, D. A., Molton, P. M., Russell, J. A., Hallen, R. T. 1984. Application of direct thermal liquefaction for the conversion of cellulosic biomass. *Ind Eng Chem Prod Res Dev* 23(3):471–475.

[NIFC] National Interagency Fire Center. 2009. Fire information: National fire news. Available from: http://www.nifc.gov/fire_info/nfn.htm. Accessed May 2009.

Nitschke,W. R., Wilson, C. M. 1965. Rudolph Diesel, Pionier of the Age of Power. Norman, OK, University of Oklahoma Press.

Noureddini, H., Gao, X., Philkana, R. S. 2005. Immobilized Pseudomonas cepacia lipase for biodiesel fuel production from soybean oil. *Bioresource Technol* 96(7):769–777.

[NRCS, USDA] National Resources Conservation Service, U.S. Department of Agriculture. 2009. Conservation showcase. Available from: http://www.nm.nrcs.usda.gov/news/showcase/showcase.html. Accessed May 2009.

[NREL] National Renewable Energy Laboratory. 2005. Biomass resources available in the United States. Available from: http://www.nrel.gov/gis/biomass.html Accessed April 2009.

NREL. 2009a. http://www.nrel.gov/otec/achievements.html. Retrieved May 12, 2009.

NREL. 2009b. http://www.nrel.gov/gis/solar.htmlAccessed April 2009.

[OECD] Organization for Economic Co-operation & Development. 2008. Directorate on trade and agriculture: Economic assessment of biofuel support policies. Paris, OECD Publishing.

Ohgren, K., Bengtsson, O., Gorwa-Grauslund, M. F., Galbe, M., Hahn-Hagerdal, B., Zacchi, G. 2006. Simultaneous saccharification and co-fermentation of glucose and xylose in steam-pretreated corn stover at high fiber content with Saccharomyces cerevisiae TMB3400. *J Biotechnol* 126(4):488–498.

Oliveira, A. J., Ramalho, J. 2006. Brazilian Agroenergy Plan 2006–2011. Ministry of Agriculture, Livestock, and Food Supply. Brasılia, Embrapa Publishing House.

Openshaw, K. 2000. A review of Jatropha curcas: An oil plant of unfulfilled promise. *Biomass Bioenerg* 19:1–15.

[ORNL] Oak Ridge National Laboratory. 2000. Boosting bioenergy and carbon storage in green plants. Available from: http://www.ornl.gov/info/ornlreview/v33_2_00/bioenergy.htm.

ORNL. 2005. Biomass as feedstock for a bioenergy and bioproducts industry: The technical feasibility of a billion-ton annual supply. Available from: http://feedstockreview.ornl.gov/pdf/billion_ton_vision.pdf. Accessed April 2005.

ORNL. 2009. Biochar. Available from: http://bioenergy.ornl.gov/papers/misc/biochar_factsheet.html. Accessed May 2009.

Osato, K., Omura, M., Suto, Y., et al. 2004. High-pressure treatment apparatus and method of operating high-pressure treatment apparatus. PCT Int. Appl., 19, pp., CODEN: PIXXD2 WO 2004105927 A2 20041209.

Overend, R. P. 1996. Production of electricity from biomass crops: US perspective. Golden, CO, National Renewable Energy Laboratory.

Papadikis, K., Gu, S., Bridgwater, A. V. 2009. CFD modelling of the fast pyrolysis of biomass in fluidised bed reactors: Modelling the impact of biomass shrinkage. *Chem Eng J* 149(1–3):417–427.

Patzlaff, J., Liu, Y., Graffmann, C., Gaube, J. 1999. Studies on product distributions of iron and cobalt catalyzed Fischer–Tropsch synthesis. *Appl Catal. A-Gen* 186(1–2):109–119.

Penche, C. 1998. Layman's Guidebook on How to Develop a Small Hydro Site. European Small Hydropower Association (ESHA), Directorate General for Energy (DG XVII).

Perez-Verdin, G., Grebner, D. L., Munn, I. A., Sun, C., Grado, S. C. 2008. Economic impacts of woody biomass utilization for bioenergy in Mississippi. *Forest Prod J* 58(11):75–83.

Peterson, A. A., Vogel, F., Lachance, R. P., Froling, M., Antal, M. J., Jr., Tester, J. W. 2008. Thermochemical biofuel production in hydrothermal media: A review of sub- and supercritical water technologies. *Energ Environ Sci* 1:32–65.

Petit, J. R., Jouzel, J., Raynaud, D., et al. 1999. Climate and atmospheric history of the past 420,000 years from the Vostok ice core, Antarctica. *Nature* 399(6735):429–436.

Petrou, E. C., Pappis, C. P. 2009. Biofuels: A survey on pros and cons. *Energ Fuel* 23(2):1055–1066.

Phillip, E. S. 1999. Organic chemical reactions in supercritical water. *Chem Rev* 99:603–621.

Pinzi, S., Garcia, I. L., Lopez-Gimenez, F. J., Luque de Castro, M. D., Dorado, G., Dorado, M. P. 2009. The ideal vegetable oil-based biodiesel composition: A review of social, economical and technical implications. *Energ Fuel* 23(5):2325–2341.

Pirkle, J. L., Kaufmann, R. B., Brody, D. J., et al. 1998. Exposure of the U.S. population to lead, 1991–1994. *Environ Health Persp* 106:745–750.

Prins, M. J., Ptasinski, K. J., Janssen, F. J. J. G. 2004. Exergetic optimisation of a production process of Fischer–Tropsch fuels from biomass. *Fuel Proc Technol* 86:375–389.

Ragauskas, A. J., Williams, C. K., Davison, B. H., et al. 2006. The path forward for biofuels and biomaterials. *Science* 311(5760):484–489.

Ramage, J., Scurlock, J. 1996. Biomass. In: Boyle, G., editor. Renewable Energy: Power for a Sustainable Future. Oxford, Oxford University Press, p 137–182.

Reed, T. B., Lerner, R. M. 1973. Methanol. Versatile fuel for immediate use. *Science* (Washington, DC, United States) 182:1299–1304.

[RFA] Renewable Fuels Association. 2009. Ethanol Industry Statistics, Washington, DC.

Riedel, T., Claeys, M., Schulz, H., et al. 1999. Comparative study of FTS with H_2/CO and H_2/CO_2 syngas using Fe and Co catalysts. *Appl Catal A-Gen* 186:201–213.

Robins, W. K., Hsu, C. S. 2000. Petroleum composition. In: Kirk-Othmer Encyclopedia of Chemical Technology. Hoboken, NJ, John Wiley & Sons.

Saga, K., Yokoyama, S., Imou, K., Kaizu, Y. 2008. A comparative study of the effect of CO_2 emission reduction by several bioenergy production systems. *Int Energy J* 9:53–60.

Saka, S., Kusdiana, D. 2001. Biodiesel fuel from rapeseed oil as prepared in supercritical methanol. *Fuel* 80:225–231.

Salmenoja, K. 1993. Black-liquor gasification: Theoretical and experimental studies. *Bioresource Technol* 46:167–171.

Sanderson, M. A., Adler, P. R. 2008. Perennial forages as second generation bioenergy crops. *Int J Mol Sci* 9:768–788.

Santos, D. T., Sarrouh, B. F., Rivaldi, J. D., Converti, A., Silva, S. S. 2008. Use of sugarcane bagasse as biomaterial for cell immobilization for xylitol production. *J Food Eng* 86:542–548.

Sasaki, M., Goto, K., Tajima, K., Adschiri, T., Arai, K. 2002. Rapid and selective retro-aldol condensation of glucose to glycolaldehyde in supercritical water. *Green Chem* 4:285–287.

Sauve, S., Mcbride, M. B., Hendershot, W. H. 1997. Speciation of lead in contaminated soils. *Environ Pollut* 98:149–155.

Savoie, P., Descoteaux, S. 2004. Artificial drying of corn stover in mid-size bales. *Can Biosys Eng* 46:225–226.

Schmitz, T. G., Seale, J. L., Buzzanell, P. 2002. Brazil's domination of the world sugar market. In: Schmitz, A., Spreen, T. H., Messina, W. A., Jr., Moss, C. B., editors. Sugar and Related Sweetener Markets: International Perspectives. Oxfordshire, CABI Publishing, pp. 123–139.

Schulz, H. 1999. Short history and present trends of FT synthesis. *Appl Catal A-Gen* 186:1–16.

Scott, D. S. 1988. Pyrolysis process for biomass. Canadian Patent No. 1241541, September.

Shaddix, C. R., Hardesty, D. R. 1999. Combustion properties of biomass flash pyrolysis oils: Final project report, Sandia Report (SAND99–8238) prepared by Sandia National Laboratories, California.

Shah, S., Sharma, S., Gupta, M. N. 2004. Biodiesel preparation by lipase-catalyzed transesterification of Jatropha oil. *Energ Fuel* 18:154–159.

Sheehan, J., Dunahay, T., Benemann, J., Roessler, P. 1998. A Look Back at the U.S. Department of Energy's Aquatic Species Program – Biodiesel from Algae. Golden, CO, National Renewable Energy Laboratory (NREL) Report: NREL/TP-580-24190.

Sierra, R., Smith, A., Granda, C., Holtzapple, M. T. 2008. Producing fuels and chemicals from lignocellulosic biomass. *Chem Eng Prog* 104(8):S10–S18.

Sievers, C., Valenzuela-Olarte, M. B., Marzialetti, T., Musin, I., Agrawal, P. K., Jones, C. W. 2009. Ionic-liquid-phase hydrolysis of pine wood. *Ind Eng Chem Res* 48(3):1277–1286.

Sims, R. E. H. 2002. The Brilliance of Bioenergy: In Business and in Practice. London, Earthscan Publications.

Smeets, E. M. W., Faaij, A. P. C. 2007. Bioenergy potentials from forestry in 2050. An assessment of the drivers that determine the potential. *Climatic Change* 81:353–390.

Spath, P. L., Dayton, D. C. 2003. Preliminary screening: Technical and economic assessment of synthesis gas to fuels and chemicals with emphasis on the potential for biomass-derived syngas. NREL/TP-510–34929, December.

Speidel, H. K., Lightner, R. L., Ahmed, I. 2000. Biodegradability of new engineered fuels compared to conventional petroleum fuels and alternative fuels in current use. *Appl Biochem Biotechnol* 84–86:879–897.

Srivastava, A., Prasad, R. 2000. Triglycerides-based diesel fuels. *Renew Sust Energ Rev* 4:111–133.

Stelmachowski, M., Nowicki, L. 2003. Fuel from the synthesis gas-the role of process engineering. *Appl Energ* 74:85–93.

Steynberg, A. P., Dry, M. E., Davis, B. H., Breman, B. B. 2004. Fischer–Tropsch reactors. In: Steynberg, A., Dry, M., editors. Fischer–Tropsch Technology. Amsterdam, Elsevier, pp. 64–96.

Sumathi, S., Chai, S. P., Mohamed, A. R. 2008. Utilization of oil palm as a source of renewable energy in Malaysia. *Renew Sust Energ Rev* 12(9):2404–2421.

Swenson, D. 2008. The economic impact of ethanol production in Iowa. Ames, IA, Iowa State University. Available from http://www.econ.iastate.edu/research/webpapers/paper_12865.pdf Accessed March 5, 2008.

Taylor, R. W., Berjoan, R., Coutures, J. P. 1980. Solar gasification of carbonaceous materials. Report No. UCRL-53063, Lawrence Livermore Laboratory. Livermore, CA, California University.

Theander, O. 1985. Cellulose, hemicellulose, and extractives. In: Overand, R. P., Mile, A. T., Mudge, L. K., editors. Fundamentals of Thermochemical Biomass Conversion. London, Elsevier, pp. 35–60.

Tijmensen, M. J. A., Faaij, A. P. C., Hamelinck, C. N., van Hardeveld, M. R. M. 2002. Exploration of the possibilities for production of Fischer Tropsch liquids and power via biomass gasification. *Biomass Bioenerg* 23:129–152.

Timell, T. E. 1967. Recent progress in the chemistry of wood hemicelluloses. *Wood Sci Technol* 1(1):45–70.

Trieb, F. 2000. Competitive solar thermal power stations until 2010: The challenge of market introduction. *Renew Energ* 19:163–171.

Tyner, W. E., Taheripour, F. 2007. Renewable energy policy alternatives for the future. *Am J Agr Econ* 89:1303–1310.

[UCS] Union of Concerned Scientists. 2009. Heat-trapping gases. Available from: http://www.ucsusa.org/publications/catalyst/heat-trapping-gasses.html. Accessed August 2009.

United Nations. 2006. The emerging biofuels market: Regulatory, trade and development implications. United Nations Conference on Trade and Development, New York and Geneva.

[UNDP] United Nations Development Programme. 2000. World Energy Assessment. Energy and the challenge of sustainability.

[UNEP] United Nations Environment Programme. 2007. Global Environment Outlook (GEO-4): Environment for development. UNEP, Kenya.

UNEP. 2008. UNEP/DEWA/GRID-Europe, GEO Data Portal. Compiled from CDIAC, Marland.G. T.A. Boden, and R.J. Anders. 2008. Global, Regional, and National Fossil Fuel CO2 Emissions.

University of Cambridge. 2009. The history behind the ozone hole. Available from: http://www.atm.ch.cam.ac.uk/tour/part1.html. Accessed July 2009.

U.S. Congress. 2007. Energy Independence and Security Act of 2007.

[USDA] United States Department of Agriculture. 2003. Production Estimates and Crop Assessment Division Foreign Agricultural Service. EU: Biodiesel Industry Expanding Use of Oilseeds. Available from: http://www.fas.usda.gov/pecad2/highlights/2003/09/biodiesel3/ USDA. 2006. The economic feasibility of ethanol production from sugar in the United States. Washington DC, July. Available from: http://www.usda.gov/oce/reports/energy/EthanolSugarFeasibilityReport3.pdf.

[U.S. DOE] U.S. Department of Energy. 2006. Biodiesel handling and use guidelines. DOE / GO-102006–2358, Third Edition, Oak Ridge, TN.

U.S. DOE. 2008. Lawrence Livermore National Laboratory, Available from: https://eed.llnl.gov. Accessed March 11, 2009.

U.S. DOE. 2009. Energy KIDS. Available from: http://www.eia.doe.gov/kids/energyfacts. Accessed March 11, 2009.

U.S. DOE, Energy Efficiency and Renewable Energy (EFRE). 2009. Ethanol myths and facts. Available from: http://www1.eere.energy.gov/biomass/printable_versions/ethanol_myths_facts.html. Accessed August 2009.

[U.S. EPA] U.S. Environmental Protection Agency. 2002. A comprehensive analysis of biodiesel impacts on exhaust emissions. Draft Technical Report, EPA420-P-02–001, October.

U.S. EPA. 2009a. Available from: http://www.epa.gov/air/airtrends/2007/graphics/Air_pollution_pathways_textbox.gif. Accessed April 2009.

U.S. EPA. 2009b. Nitrogen oxides. Available from: http://www.epa.gov/air/emissions/nox.htm. Accessed April 2009.

U.S. EPA. 2009c. Sulfur oxide. Available from: http://www.epa.gov/air/emissions/so2.htm. Accessed April 2009.

U.S. EPA. 2009d. Mercury modeling in watersheds and water bodies. Available from: http://www.epa.gov/athens/research/modeling/mercury./$<\tau\pi/>$ Accessed April 2009.

Utlu, Z. 2007. Evaluation of biodiesel obtained from waste cooking oil. *Energy Sources* 29:1295–1304.

Van, G. J., Shanks, B., Pruszko, R., Clements, D., Knothe, G. 2004. Biodiesel Production Technology. Golden, CO, National Renewable Energy Laboratory. Paper Contract No.: DE-AC36–99-GO10337.

Van Steen, E., Claeys, M. 2008. Fischer-Tropsch catalysts for the biomass-to-liquid process. *Chem Eng Technol* 31(5):655–666.

Vidal, B. J. W. 2006. A photosynthesis civilization II (A Civilizac-ão da Fotossíntese II). See also: http://www.institutodosol.org.br/artigos.asp#S.

Wagner, W., Pruss, A. 2002. The IAPWS Formulation 1995 for the thermodynamic properties of ordinary water substance for general and scientific use. *J Phys Chem Ref Data* 31(2):387–535.

Walker, J. D., Petrakis, L., Colwell, R. R. 1976. Comparison of biodegradability of crude and fuel oils. *Can J Microbiol* 22:598–602.

Walter, A., Cortez, L. 1999. An historical overview of the Brazilian bioethanol program. *Renew Energ Dev* 11:2–4.

Walters, C. C. 2006. The origin of petroleum. *Practical Advances in Petroleum Processing* 1:79–101.

Wardle, D. A. 2003. Global sale of green air travel supported using biodiesel. *Renew Sust Energ Rev* 7(1):1–64.

[WEC] World Energy Council. 2004. Survey of energy resources. London, WEC.

Wenzl, H. F. J. 1970. The Chemical Technology of Wod. New York, NY, Academic Press.

White, D. H., Wolf, D. 1987. A continuous extruder-feeder for reactor systems for biomass fuels processing. *Energ Biomass Wastes* 10:1685–1688.

White, D. H., Wolf, D. 1995. Direct biomass liquefaction by an extruder-feeder system. *Chem Eng Commun* 135: 1–19.

[WHO] World Health Organization. 2005. Air quality guidelines for particulate matter, ozone, nitrogen dioxide and sulfur dioxide.WHO, Switzerland.

Williams, R. H., Larson, E. D. 1996. Biomass gasifier gas turbine power generating technology. *Biomass Bioenerg* 10(2–3):149–166.

Witze, A. 2007. Energy: That's oil, folks … *Nature* 445:14–17.

Wolfson, A., Litvak, G., Dlugy, C., Shotland, Y., Tavor, D. 2009. Employing crude glycerol from biodiesel production as an alternative green reaction medium. *Ind Crop Prod* 30(1):78–81.

Wu, B. S., Bai, L., Xiang, H. W., Li, Y. W., Zhang, Z. X., Zhong, B. 2004. An active iron catalyst containing sulfur for Fischer–Tropsch synthesis. *Fuel* 83:205–512.

Yamaguchi, T. 1998. Structure of subcritical and supercritical hydrogen-bonded liquids and solutions. *J Mol Liq* 78:43–50.

Yang, H., Zhou, Y., Liu, J. 2009. Land and water requirements of biofuel and implications for food supply and the environment in China. *Energ Policy* 37:1876–1885.

Yesodharan, S. 2002. Supercritical water oxidation: An environmentally safe method for the disposal of organic wastes. *Curr Sci* 82(9–10):1112–1122.

Yu, Y., Lou, X., Wu, H. W. 2008. Some recent advances in hydrolysis of biomass in hot-compressed water and its comparisons with other hydrolysis methods. *Energ Fuel* 22(1):46–60.

Yudovich, Ya. E., Ketris, M. P. 2005. Mercury in coal: A review. Part 1. Geochemistry. *Int J Coal Geol* 62(3):107–134.

Yung, M. M., Jablonski, W. S., Magrini-Bair, K. A. 2009. Review of catalytic conditioning of biomass-derived syngas. *Energ Fuel* 23(4):1874–1887.

Zhang, B., Keitz, M., Valentas, K. 2008. Thermal effects on hydrothermal biomass liquefaction. *Appl Biochem Biotechnol* 147:143–150.

Zhang, Q., Chang, J., Wang, T. J., Xu, Y. 2007. Review of biomass pyrolysis oil properties and upgrading research. *Energ Convers Manage* 48:87–92.

Zhang, X., Peterson, C., Reece, D., Haws, R., Moller, G. 1998. Biodegradability of biodiesel in the aquatic environment. *Transactions of the ASAE* 41:1423–1430.

Zhang, Y., Dub, M. A., McLean, D. D., Kates, M. 2003. Biodiesel production from waste cooking oil. 2. Economic assessment and sensitivity analysis. *BioresourceTechnol* 90:229–240.

Zhao, C., Kou, Y., Lemonidou, A. A., Li, X. B., Lercher, J. A. 2009. Highly selective catalytic conversion of phenolic bio-oil to alkanes. *Angew Chem Int Ed Engl* 48(22):3987–3990, S3987/1–S3987/6.

Zhu, L., O'Dwyer, J. P., Chang, V. S., Granda, C. B., Holtzapple, M. T. 2008. Structural features affecting biomass enzymatic digestibility. *Biores Technol* 99(9):3817–3828.

Index

LaVergne, TN USA
03 February 2011
214870LV00001B/123-306/P

9 780521 763998